MATERIALS RESEARCH SOCIETY
SYMPOSIUM PROCEEDINGS VOLUME 1031

Nanostructured Solar Cells

November 26-30, 2007
Boston, Massachusetts, USA

Printed from e-media with permission by:

Curran Associates, Inc.
57 Morehouse Lane
Red Hook, NY 12571
www.proceedings.com

ISBN: 978-1-60560-824-2

Some format issues inherent in the e-media version may also appear in this print version.

CAMBRIDGE UNIVERSITY PRESS
Cambridge, New York, Melbourne, Madrid, Cape Town,
Singapore, São Paulo, Delhi, Tokyo, Mexico City

Cambridge University Press
32 Avenue of the Americas, New York, NY 10013-2473, USA

www.cambridge.org

Materials Research Society
506 Keystone Drive, Warrendale, PA 15086
http://www.mrs.org

©Materials Research Society 2008

This publication is in copyright. Subject to statutory exception
and to the provisions of relevant collective licensing agreements,
no reproduction of any part may take place without the written
permission of Cambridge University Press.

First published 2008

CODEN: MRSPDH

ISBN: 978-1-60560-824-2

Cambridge University Press has no responsibility for the persistence or
accuracy of URLs for external or third-part Internet Web sites referred to
in this publication and does not guarantee that any content on such Web sites
is, or will remain, accurate or appropriate.

Additional copies of this publication are available from:

Curran Associates, Inc.
57 Morehouse Lane
Red Hook, NY 12571 USA
Phone: 845-758-0400
Fax: 845-758-2634
Email: curran@proceedings.com
Web: www.proceedings.com

TABLE OF CONTENTS

Thermal Dependence of Quantum Dot Solar Cells 1
C. D. Cress, S. M. Hubbard, C. Bailey, R. Robinson, B. J. Landi, R. P. Raffaelle

Electrophoretic Deposition of CdSe Nanocrystals for Photovoltaic Applications 7
N. J. Smith, K. J. Emmett, S. J. Rosenthal

Fabrication of Copper Indium Diselenide Nanowires 13
S. Phok, S. Rajaputra, V. Singh

Hierarchical Assembly of 2D Nanostructures of Relevance for Organic Solar Cell Design 19
S. O'Donnell, M. Buettner, P. Reinke

Doped Silicon Nanoparticles Synthesized by Nonthermal Plasma 25
X. Pi, R. Anthony, S. Campbell, U. Kortshagen

**III-V Semiconductor Vertical and Tilted Nanowires on Silicon Using Chemical Beam
Epitaxy** 31
G. Radhakrishnan, A. Freundlich, J. Charlson, B. Fuhrmann

Analysis of Strain Compensation in Quantum Dot Embedded GaAs Solar Cells 37
C. Bailey, C. Cress, R. Raffaelle, S. Hubbard, W. Maurer, D. Wilt, S. Bailey

**Solidification -Fabrication of Charge Carrier Path Solidification -Fabrication of Charge
Carrier Path** 43
F. Inakazu, Y. Ogomi, Y. Noma, Y. Fujita, M. Kono, Y. Yamaguchi, Y. Kashiwa, T. Kogo, S. Hayase

Polaron Pair Dissociation and Polaron Recombination in Polymer: Fullerene Solar Cells 49
C. Deibel, A. Baumann, J. Lorrmann, V. Dyakonov

Environmental Passivation and Temperature Cycling of PCBM - Polymer Solar Cells 55
A. Anctil, A. Merrill, C. Cress, B. Landi, R. Raffaelle

Rigid-Rod Sensitizers bound to Semiconductor Nanoparticles 61
O. Taratula, E. Galoppini

**Determination of Nanomaterial Energy Levels for Organic Photovoltaics by Cyclic
Voltammetry** 67
R. A. DiLeo, A. Anctil, B. Landi, C. Cress, R. P. Raffaelle

**Synthetic Approaches to Study Aggregation of Tripodal Linkers on Semiconductor
Surfaces** 73
S. Thyagarajan, E. Galoppini

**Energy Band Engineering for Improved Vertical Transport in Quantum Structured III-V
p-i-n Solar Cells** 79
A. Alemu, A. Freundlich

Evidence of Sequential Carrier Escape in III-V p-i-n Multi-Quantum Well Solar Cells 85
A. Alemu, J. A. H. Coaquira, A. Freundlich

Efficient Thin Polymer Solar Cells with Post-Annealing 91
S. W. Liu, C. C. Lee, P. T. Huang, C. T. Chen, J. K. Wang

Copper Phthalocyanine Nanowire Based Solar Cells 96
V. Singh, S. Rajaputra, S. Phok, G. Chintakula, G. Sagi

Advances in the Research of the Intermediate Band (IB) Solar Cell 102
A. Luque, A. Marti

Author Index

Thermal Dependence of Quantum Dot Solar Cells

Cory D. Cress[1], Seth M. Hubbard[1,2], Christopher Bailey[1], Ross Robinson[1], Brian J. Landi[1], and Ryne P. Raffaelle[1,2]

[1]NanoPower Research Laboratories, Rochester Institute of Technology, 85 Lomb Memorial Drive, Rochester, NY, 14623
[2]Physics, Rochester Institute of Technology, 85 Lomb Memorial Drive, Rochester, NY, 14623

ABSTRACT

Various temperature dependent optoelectronic properties were measured for GaAs-based p-type / intrinsic / n-type (*pin*) solar cell devices containing 5-layers of InAs quantum dots (QDs) grown with strain-compensation layers. Curve fitting of the dark diode characteristics allowed for the temperature dependence of the saturation current and the ideality parameter to be determined. The resulting parameter values indicate high material quality. Air mass zero illuminated current density vs. voltage measurements were used to determine the temperature coefficients of the open circuit voltage, short circuit current, maximum power, and fill factor. A strong correlation between the temperature dependent quantum dot electroluminescence peak emission wavelength and the sub-GaAs bandgap spectral responsivity was observed.

INTRODUCTION

Many recent experimental investigations have demonstrated the ability to enhance the spectral responsivity of a GaAs solar cell by incorporating either InAs QDs [1-4] or GaSb QD [5]. In devices which utilized the stranski-krastanow QD growth mode, an improved long-wavelength responsivity has been achieved, although a concomitant reduction in short circuit current is also observed. The reduction in short circuit current has been attributed to strain-induced emitter degradation [4]. Electron trapping by the defects at the QD / GaAs interface, which increases the recombination current, may also contribute to the decreased performance [6]. As additional InAs QDs layers are grown using the Stranski-Kranstanow growth mode, residual compressive strain accumulates in the structure. To alleviate this strain in QD laser diodes [7, 8] and multiple-quantum well solar cells [9], a layer of material under tensile strain is grown to offset the compressive strain leading to a strain-neutral stack. Promising results have recently been demonstrated in GaAs-based InAs QD solar cells using this strain compensation technique, although the thickness of the strain compensation layers was still under investigation [10].

In this paper, the thermal behavior of a strain-compensated QD solar cell with a room temperature air mass zero efficiency in excess of 13% is reported. The thermal behavior of QD solar cells is an important characteristic that must be evaluated to determine the suitability of these devices for space and terrestrial concentrator applications where high temperatures may be encountered. In space, the operating temperature can vary drastically as the solar cells pass into and out of solar eclipse. Likewise, solar cells used in terrestrial concentrator applications are expected to be operated at 250 suns or higher and typically rely solely on passive cooling systems [11]. Measuring the temperature dependent optoelectronic and photovoltaic behavior of QD solar cells, therefore, will help to determine their suitability for these applications and may offer additional insight into the fundamental role of the QDs in these devices.

EXPERIMENT

The devices invested in this study consist of a strain-compensated GaAs *pin* device and a reference GaAs device both 1 cm^2 in area and containing a 4% grid shadowing. The strain compensated device (referred to as 5x QD) is comprised of 5-layers of InAs QDs, each separated by a 10 nm intrinsic GaAs cladding layer and a GaP strain relief layer, grown within the middle of the intrinsic region. The device structure along with the metal organic vapor phase epitaxy (MOVPE) growth conditions have been reported elsewhere [10]. The temperature of the devices was controlled for the dark and illuminated current density vs. voltage (*J-V*), spectral responsivity (*SR*), and electroluminescence (*EL*) measurements, by an Oxford Instruments cryostage affixed with a quartz window. The temperature was varied in 20°C increments over the temperature range of 100 K – 420 K. Full illumination of the devices was used in the *J-V* and *SR* measurements and the band pass of the quartz window was ~90% over the wavelength range of 350 nm - 2200 nm. A Newport Oriel solar simulator affixed with air mass zero filtering and a 1000 W xenon bulb was used as the broad-band illumination source. An Optronic Laboratories' OL 750 Spectroradiometer was used to measure the monochromatic spectral responsivity of the devices. To measure the electroluminescence spectra, the emission from the devices was chopped (167 Hz) and focused onto the entrance slit of a 1.26 m path-length monochromator. The chopped signal was detected using a thermoelectrically cooled InGaAs photodiode and measured using a Stanford Research Systems lock-in amplifier. The devices were forward biased with a constant current of 200 mA/cm^2.

DISCUSSION

The dark diode response of the reference *pin* device (referred to as PIN) and the strain compensated QD solar cell (5x QD) are depicted in Figure 1a and b, respectively. The strong temperature dependence observed in both devices is characteristic of injection and space charge region recombination currents; tunneling current, an additional forward current mechanism, is largely temperature independent. Injection and space charge region recombination both display a Boltzmann-type thermal dependence wherein the injection current and space charge region recombination current vary in relation to the intrinsic carrier concentration as n_i^2 and n_i respectively [12]. Based on the value of the ideality parameters shown in Figure 1c, the PIN device suffers from space charge region recombination although the saturation currents (see Figure 1d) are quite low indicating high quality material. Although the 5x QD device has a wide depletion region (*i.e.*, 0.1 μm intrinsic region including 5-layers of InAs QDs and GaP) it still achieved an ideality parameter of ~ 1.5 at room temperature. In comparison to the PIN device, improved idealities and lower saturation currents are observed in the 5x QD device. The improved device performance may result from the GaP layers grown within the intrinsic region. With a bulk bandgap of 2.26 eV, such layers have a comparably lower intrinsic carrier concentration and a correspondingly lower contribution to the dark current. For intrinsic GaAs and GaP materials, the offsets in ionization potential leads to GaP conduction and valence bands which straddle that of the adjacent GaAs layers. This may hinder the transport of holes into the base and electrons into the emitter since the band offsets result in additional energy barriers for the charge carriers to overcome [13]. However, the improved performed observed may be a consequence of greater confinement of the QD states due to the wider GaP energy band gap. This results in a band structure which is similar to that of the DFENCE heterostructure proposed in ref. [14], the difference being the thin GaAs capping layers grown between the QDs and the GaP strain compensation layers.

The temperature dependent *J-V* characteristics under simulated air mass zero (AM0) illumination are depicted in Figure 2a and b for the PIN and 5x QD devices, respectively. At room temperature, both devices have comparable short circuit currents densities (J_{sc}), while the open circuit voltage (V_{oc}) is slightly greater in the PIN device. Figure 2c illustrates the temperature evolution of the J_{sc}, V_{oc}, fill factor (*FF*), and efficiency (η), for the two devices under AM0. It is apparent that most of the parameters vary linearly with temperature although the J_{sc} has a slight variation between low and high temperature operation.

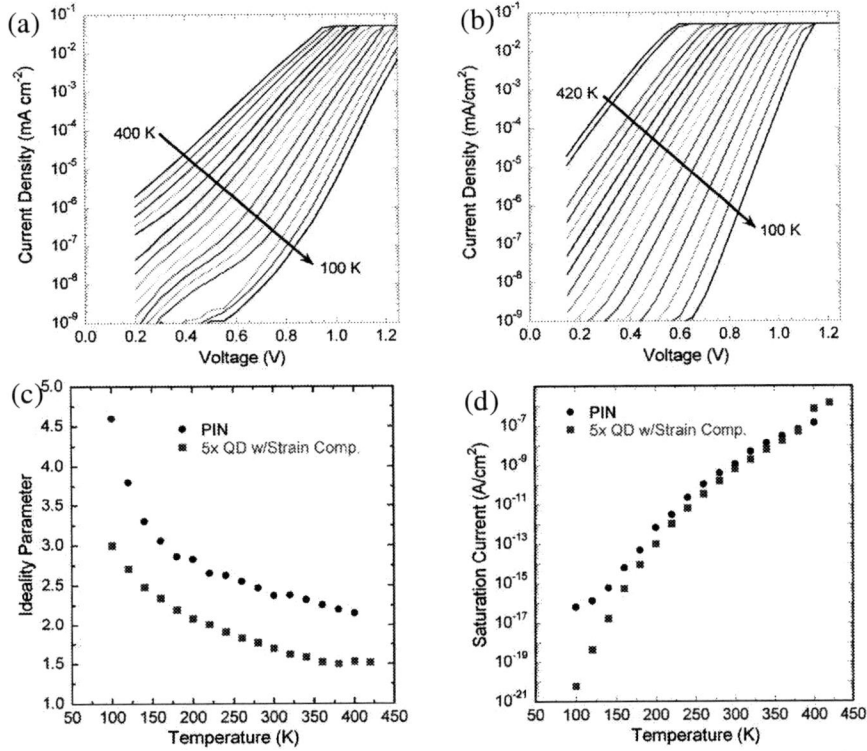

Figure 1. The temperature dependent dark *J-V* characteristics for the (a) PIN device and (b) 5x QD device. In (c) and (d) the ideality parameter and the saturation current, extracted from dark diode characteristics and based on the ideal diode equation, are plotted for both devices as a function of temperature, respectively.

The temperature coefficients were obtained from the slopes of linear fits (assuming an ideal diode) for the various parameters, over the specified temperature ranges (two ranges for the J_{sc}), and are provided in Table I. The V_{oc} dependence on temperature (see Figure 2c) are comparable in both devices, the 5x QD being slightly greater in magnitude. The V_{oc} temperature dependence of GaAs-based devices typically decreases linearly with temperature and is primarily due to increases in dark current [12]. Referring back to Figure 1d, the saturation current increases at a slightly greater rate in the 5x QD device which corresponds to the slightly greater V_{oc} temperature coefficient observed in the 5x QD device. A positive I_{sc} temperature coefficient is commonly observed in GaAs-based solar cells [12], and is also observed in the data depicted in Figure 2c. This is the result of increased absorption due to bandgap narrowing with temperature. In the high temperature range, the 5x QD device has a slightly greater temperature coefficient than the PIN device which may result from improved junction transport at elevated temperatures (*i.e.*, easier to overcome GaP barriers).

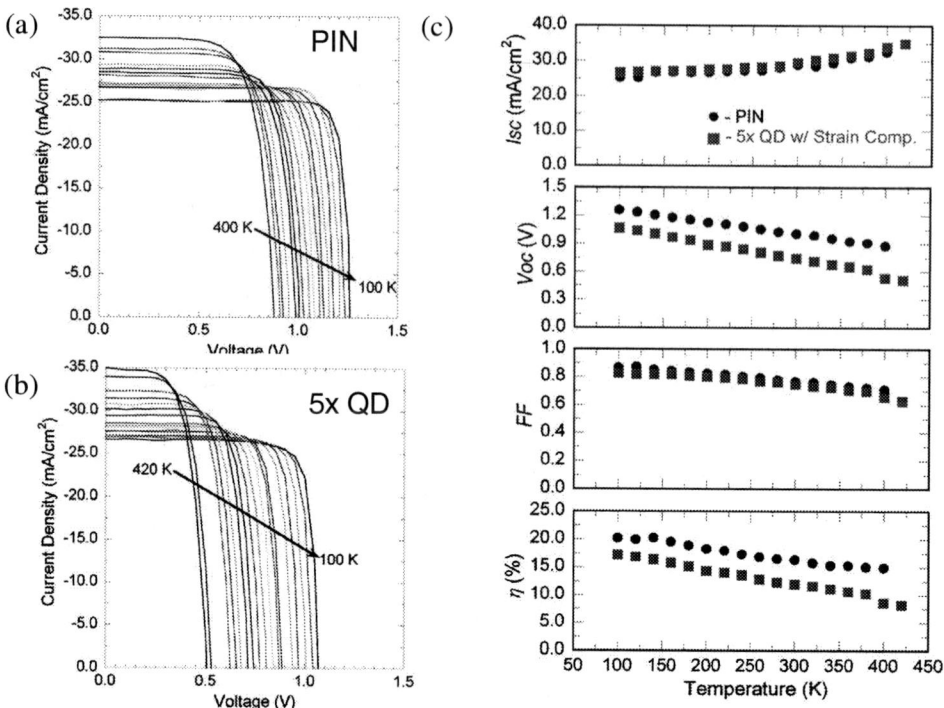

Figure 2. Temperature dependent *J-V* characteristics of (a) PIN and (b) 5x QD under simulated air mass zero illumination. (c) The short circuit current, open circuit voltage, fill factor, and efficiency of the two devices over the temperature range of 100 K – 420 K.

A decreasing *FF* with temperature, as shown in Figure 2c, has been observed in GaAs-based devices in the past [12]. In the devices investigated here, the *FF* temperature coefficients are nearly identical. Although the I_{sc} increases with temperature, the decreased V_{oc} and *FF* offset this effect yielding a net negative efficiency temperature coefficient. These values are consistent with those obtained for the strain compensated devices investigated here, which indicates that the incorporation of the InAs QDs and the GaP strain compensation layers do not have any substantial detrimental effects on the temperature dependence of the device.

Table I. Temperature coefficients for the two devices.

Temperature Coefficient	PIN	5x QD	Temperature Range
$\Delta I_{sc}/\Delta T$ ($\mu A \cdot cm^{-2} \cdot K^{-1}$)	13.4	9.4	100-240
$\Delta I_{sc}/\Delta T$ ($\mu A \cdot cm^{-2} \cdot K^{-1}$)	31.5	34.9	260-380
$\Delta V_{oc}/\Delta T$ ($mV \cdot K^{-1}$)	-1.26	-1.61	100-380
$\Delta FF/\Delta T$ ($\times 10^{-3} \cdot K^{-1}$)	-0.56	-0.55	160-380
$\Delta \eta/\Delta T$ ($\times 10^{-3} \cdot K^{-1}$)	-20	-25	160-380
$\Delta \eta/\Delta T$ ($\times 10^{-3} \cdot K^{-1}$)	Typical Values: -20 to -30		Ref. [12]

Figure 3a and b contain the spectral responsivity of the two devices measured at incremental temperatures. The spectral responsivities of the two devices are nearly identical throughout the wavelength range that corresponds to photons of energy greater than that of the GaAs bandgap. However, in the sub-GaAs bandgap region (subgap) the InAs photogeneration is clearly visible in the 5x QD device. The spectral responsivity over the subgap range shows the enhancement in the responsivity by incorporating the InAs QDs (see Figure 3c). The 100 K and 420 K responsivity spectra for the PIN device envelop those of the 5x QD device in the wavelength range near the peak responsivity of the device further emphasizing the similarity in the temperature response of the two GaAs-based devices.

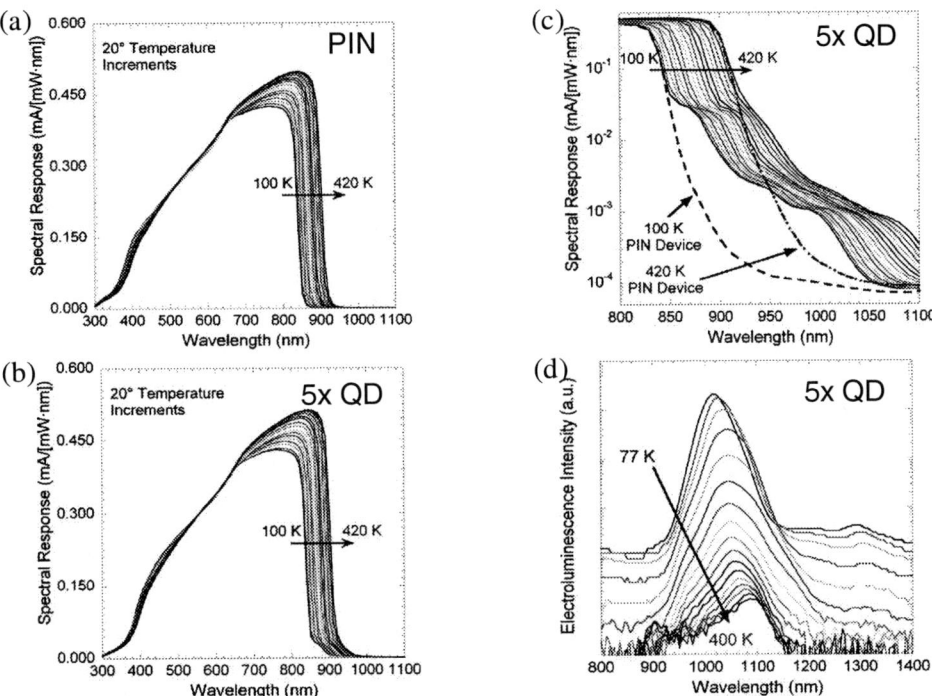

Figure 3. Spectral responsivity over the temperature range of 100 K – 420 K for (a) the PIN device and (b) the 5x QD device, respectfully. In (c), the spectral responsivity of the 5x QD device is plotted on a log-scale over the wavelength range of 800-1100 nm to better illustrate the InAs QD photogenerated current. Additionally, the 100 and 420 K responsivity spectra for the PIN device are included as dashed and dash-dot lines. (d) Electroluminescence emission spectra from the 5x QD device.

QD spectral responsivity peaks corresponding to narrow banded electronic states are observed (in the 100 K spectrum) at 875 nm, 960 nm, and 1010 nm in this device. The magnitude of these peaks does not increase with temperature, although they do shifts towards longer wavelengths. Such behavior is suggestive of a temperature independent photon-assisted carrier extraction mechanism, as opposed to a carrier thermalization extraction mechanism which would cause them to increase with temperature. Furthermore, the 1010 nm peak, corresponding to the first excited state of the InAs QDs correlates well with the peak observed in the *EL* spectra shown in Figure 3d. The narrow full width at half maximum of the *EL* data and single peak maximum indicate strong confinement and narrow size dispersion.

CONCLUSIONS

The temperature dependent optoelectronic properties were measured for a GaAs *pin* solar cell containing 5-layers of InAs quantum dots (QDs) grown with GaP strain-compensation layers. The ideality and saturation currents of the device were slightly better than a reference GaAs *pin* device which may be related to greater confinement of carriers by the wide bandgap GaP layers. The temperature coefficients of the V_{oc}, J_{sc}, η, and *ff* were compared with that of a reference GaAs *pin* device showing nearly identical thermal dependence. A strong correlation between the temperature dependent quantum dot electroluminescence peak emission wavelength and the sub-GaAs bandgap spectral responsivity is observed. In general, the temperature dependence of the QD solar cell demonstrated similar thermal behavior as that of a conventional GaAs-based *pin* device.

ACKNOWLEDMENTS

This work was supported by the DoD and the AFOSR under the contract No. FA95500610319. Additionally, C.D.C would like to acknowledge funding by NASA under a GSRP fellowship award No. NNX07AR57H.

REFERENCES

1. C. D. Cress, S. M. Hubbard, B. J. Landi, D. M. Wilt, and R. P. Raffaelle, Appl. Phys. Lett. 91, 183108/1-3 (2007).
2. S. M. Hubbard, D. M. Wilt, S. Bailey, D. Byrnes, and R. P. Raffaelle. *Proceedings of the Proc. of the World Conference on Photovoltaic Energy Conversion* (IEEE, 2006) p. 118-121.
3. A. Marti, A. Antolin, C. R. Stanley, C. D. Farmer, N. Lopez, P. Diaz, E. Canovas, P. G. Linares, and A. Luque, Phys. Rev. Lett. 9724, 247701 (2006).
4. A. Marti, N. Lopez, E. Antolin, E. Canovas, A. Luque, C.R.Stanley, C.D.Farmer, and P. Diaz, Appl. Phys. Lett. 90, 233510/1-3 (2007).
5. R. B. Laghumavarapu, A. Moscho, A. Khoshakhlagh, M. El-Emawy, L. F. Lester, and D. L. Huffaker, Appl. Phys. Lett. 90, 173125 (2007).
6. A. Luque, A. Martí, N. Lopez, E. Antolin, E. Canovas, C. Stanley, C. Farmer, and P. Diaz, J. Appl. Phys. 99, 094503/1-9 (2006).
7. P. Lever, H. H. Tan, and C. Jagadish, Appl. Phys. Lett. 95, 5710 (2004).
8. N. Nuntawong, S. Birudavolu, C. P. Hains, S. Huang, H. Xu, and D. L. Huffaker, Appl. Phys. Lett. 853050 (2004).
9. M. Mazzer, K. W. J. Barnham, I. M. Ballard, A. Bessiere, A. Ioannides, D. C. Johnson, M. C. Lynch, T. N. D. Tibbits, J. S. Roberts, G. Hill, and C. Calder, Thin Film Solids. 76, 511-512 (2006).
10. S. M. Hubbard, R. P. Raffaelle, R. Robinson, C. Bailey, D. M. Wilt, D. Wolford, W. Maurer, and S. Bailey. *Proceedings of the Proceedings of the Materials Research Society* (MRS, 2007) p. 1017E.
11. H. Cotal and R. Sherif. *Proceedings of the World Conference on Photovoltaic Energy Conversion* (IEEE, 2006) p. 845-848.
12. H. J. Hovel, *Solar cells*. Semiconductors and Semimetals, ed. R.K. Willardson and A.C. Beer. Vol. 11. 1975, New York: Academic. 166-174.
13. S. M. Sze, *Physics of Semiconductor Devices*. 2nd ed. 1981, New York: Wiley.
14. G. Wei and S. R. Forrest, Nano Lett. 71, 218-222 (2007).

Mater. Res. Soc. Symp. Proc. Vol. 1031 © 2008 Materials Research Society   1031-H13-25

Electrophoretic Deposition of CdSe Nanocrystals for Photovoltaic Applications

Nathanael J Smith[1], Kevin J Emmett[2], and Sandra J Rosenthal[1]
[1]Department of Chemistry, Vanderbilt University, Nashville, TN, 37240
[2]Department of Physics and Astronomy, Vanderbilt University, Nashville, TN, 37240

ABSTRACT

CdSe nanocrystals chemically linked to nanocrystalline titanium dioxide substrates form a promising material for nanostructured photovoltaic devices. The usual method for attaching nanocrystals to the titanium dioxide substrate is by means of a linking molecule (such as mercaptopropionic acid) or in-situ growth. In this paper, we report the use of an alternative technique, electrophoretic deposition (EPD), to directly deposit already formed CdSe nanocrystals onto the substrate. In EPD, a voltage is established between two electrodes that are immersed in a solution of nanocrystals. At room temperature, a fraction of the nanocrystals are thermally charged, and these charged nanocrystals migrate to the electrodes and adhere to the surface. A significant advantage of EPD over current techniques is the speed with which the nanocrystals are deposited: EPD takes only a few minutes, compared to the several hours required for the alternative techniques. As a proof of principle, we have fabricated initial photovoltaic devices based on electrophoretically deposited CdSe nanocrystals on a planar TiO_2 thin film.

INTRODUCTION

Interest in semiconductor nanocrystals for use in photovoltaic cells has arisen primarily for two reasons: (1) They have easily tunable band gaps[1], allowing devices based on them to be engineered to have excellent overlap with the solar spectrum, and (2) they offer the possibility of utilizing photon energy in excess of the band gap that is currently wasted as heat [2]. The recent reports of multi-exciton generation in single semiconductor nanocrystals [3,4] has generated extra interest in the second point at the present time. One of the many ways in which quantum dot based solar cells can be realized is to use the dots to sensitize a high bandgap semiconductor, such as TiO_2, in a direct analogy to dye-sensitized solar cells [5]. The nanocrystals can be attached using a chemical linking molecule, or more commonly they are grown in-situ. These techniques are time consuming. For example, to chemically link CdSe nanocrystals to TiO_2 films using mercaptopropionic acid as the linking molecule required a total of 16 hours [5]. CdS nanocrystals grown in-situ require multiple immersions in solutions of Cd and S precursors [6,7], while growth of CdSe nanocrystals by chemical solution deposition takes several hours [8,9]. We present here an alternative method for attaching nanocrystals to TiO_2, that of electrophoretic deposition (EPD). As well as taking only a few minutes to perform, EPD can give precise control over film thicknesses.

Electrophoretic deposition is a simple thin film deposition technique. The process of EPD occurs when charged particles in solution are subject to an applied electric field. The particles move under the action of the field to electrodes immersed in the solution, where they subsequently adhere to the surface. Electrophoretic deposition has found extensive use in industry for coating a variety of products (for a survey of the technique and its applications, see for example the review by Van der Biest and Vandeperre [10]). More recently, EPD has been used to deposit films of nanoparticles. For example, TiO$_2$ nanoparticles [11] and nanotubes [12], europium oxide nanocrystals [13], CdSe nanocrystals [14-17], and γ-Fe$_2$O$_3$ nanocrystals [15] have all been deposited using EPD. In this paper, we have investigated the use of EPD to form thin films of CdSe nanocrystals for photovoltaic applications. A particularly attractive feature of EPD of CdSe nanocrystals is that at room temperature, enough nanocrystals are thermally charged to allow the process to work [14]. Thus no extra preparation of the nanocrystals is necessary, beyond the normal isolation of the nanocrystals after synthesis.

EXPERIMENT

CdSe nanocrystals were synthesized according to standard pyrolitic techniques using CdO and Se:TBP (tributylphosphine) as precursors [18]. Briefly, CdO, trioctylphosphineoxide (TOPO), hexadecylamine (HDA) and dodecylphosphonic acid (DPA) were combined in a 100 mL three-neck flask. A temperature probe, bump-trap, and rubber septum were also affixed to the flask. Argon was flowed through the flask and exhausted through a 12 gauge needle inserted through the rubber septum. The CdO mixture was heated, with vigorous stirring, to a final temperature of 330 °C. At this temperature the solution turned clear and colorless, indicating the formation of cadmium phosphonate. The temperature was allowed to drop to 315 °C and the Se:TBP solution was swiftly injected into the flask via a 12 gauge needle through the rubber septum. Formation of CdSe nanocrystals caused the solution to change from clear to yellow, then to deep red. After approximately two minutes, the solution was cooled down to room temperature and the nanocrystals separated from the surfactants by precipitation and centrifugation.

Titanium dioxide thin films were prepared by spin-coating a sol-gel precursor solution onto ITO coated glass substrates. The precursor solution was made by adding 1 g of titanium (IV) ethoxide and 0.15 g of HCl to 8 g of 2-propanol [19]. After spin-coating, the films were annealed for 1 hour at 450 °C in air to form films of anatase TiO$_2$.

Electrophoretic deposition was carried out in a 10 mL beaker filled with a solution of CdSe nanocrystals in hexanes. Electrodes were connected to a high voltage power supply and lowered into the solution (after the voltage had been established between the electrodes). Current was monitored using a Keithley 2400 source-measure unit in current measure mode. The electrode spacing was approximately 2 mm, and the voltage applied to the electrodes was 500 V for all measurements reported here. The optical density of the nanocrystal solutions used for deposition was in the range of 0.1 to 0.6. The deposition times were 20 minutes, during which time the level of the nanocrystal solution was kept constant by addition of hexanes.

DISCUSSION

Electrophoretic Deposition

For photovoltaic cells, CdSe nanocrystals were deposited onto a thin film of TiO$_2$ on ITO coated glass, as described below. However, to determine that CdSe nanocrystals were actually being deposited, films were deposited onto thin films of TiO$_2$ on silicon substrates, and Rutherford backscattering spectroscopy (RBS) used to verify the presence of nanocrystals (Figure 1). Silicon substrates were used because the Cd and Se peaks cannot be easily resolved from the very large signals from indium and tin. The ratio of Cd to Se in the nanocrystals, determined from the areas of the Cd and Se peaks in Figure 1, was 1.6:1. The excess of Cd indicates that Cd precursors left over from the nanocrystal synthesis were still present in the deposition solution. In fact we have found it very difficult to completely clean up CdSe nanocrystals synthesized using CdO as the cadmium source.

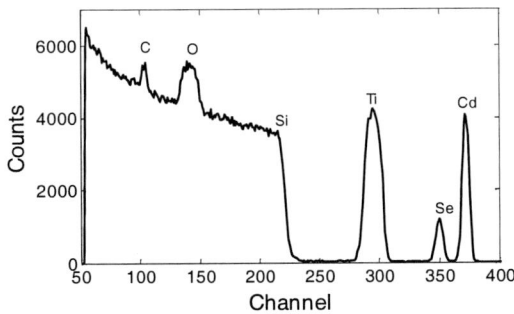

Figure 1. Rutherford backscattering spectrum of CdSe nanocrystals deposited onto a thin film of TiO_2 on a silicon substrate.

The deposition of nanocrystals using EPD was found to vary from batch to batch of nanocrystals. Three different scenarios were observed: nanocrystals were deposited on both electrodes, nanocrystals were deposited on only the positive electrode, or no nanocrystals were deposited at all. The reasons for the different behaviors from different batches of nanocrystals are not yet understood, but are likely to be related to the relative amounts of reactants left over from synthesis. As already noted, it has proved difficult to fully remove all the reactants from the nanocrystal solution, particularly HDA. Perhaps surprisingly, the current measured during all depositions, regardless of whether nanocrystals were deposited or not, was similar. This behavior has previously been reported by Islam et. al. [14].

A typical current verses deposition time trace is shown in Figure 2a. Upon insertion of the electrodes into the nanocrystal solution, the current spiked before settling into an exponential decay with time. The small spikes at 370 s and 830 s occurred when the EPD cell was topped up with hexanes. The exponential fit was obtained by fitting a single exponential decay to the data between the fast initial spike and the first "top up" spike at 370 s. For the example shown, the decay time was found to be 70 s. Interestingly, the nanocrystal concentration had little effect on this value. It appears from the figure that most of the deposition occurred in the first three to four minutes. For films deposited directly on to silicon, we found that deposition of the films was complete within the first minute. RBS spectra for films deposited for 1, 2 and 3 minutes (Figure 2b) clearly indicates that after 1 minute deposition was complete.

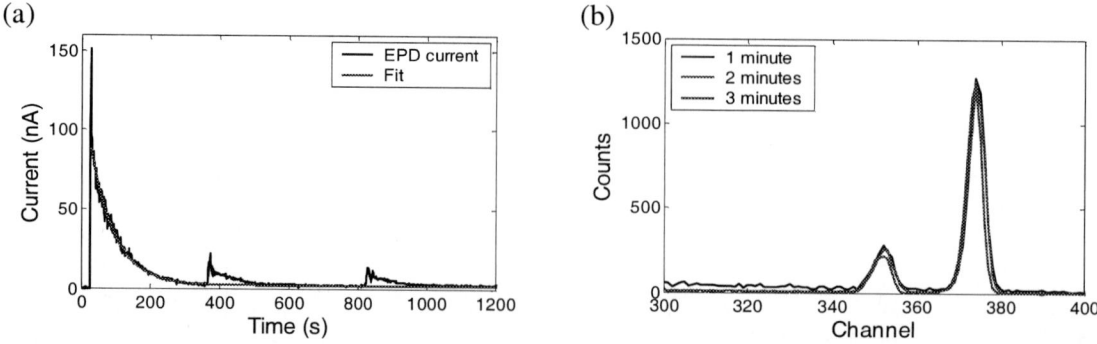

Figure 2. (a) Typical deposition current as a function of time during EPD. The current during deposition is typified by the fast current spike when the electrodes are first inserted into solution, followed by exponential decay. The small current spikes at 370 s and 830 s occurred when hexanes were added to counter evaporation. (b) RBS spectra of CdSe nanocrystals deposited directly on to silicon for 1, 2 and 3 minutes. The same amount of material was deposited in each case, irregardless of the deposition time.

Photovoltaic Cells

Photovoltaic cells were fabricated by electrophoretic deposition of CdSe nanocrystals onto a planar TiO_2 film on ITO coated glass. After deposition of the nanocrystals, a film of N,N′-Bis(3-methylphenyl)-N,N′-diphenylbenzidine (TPD), a hole transporting, conductive polymer, was deposited by spin casting, followed by a final deposition of Al contacts by thermal evaporation. The device structure is shown in Figure 3. Broadband illumination was provided by a tungsten filament lamp, focused onto the cell. The illumination intensity incident on the active area of the cells (taken to be the area of the Al electrode, area = 0.071 cm^2) was ~30 mW. Current-voltage curves under dark and light conditions were measured using a Keithley 2400 source-measure unit.

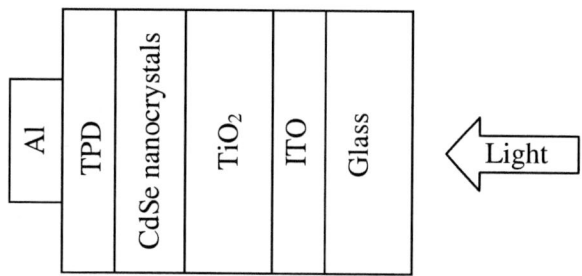

Figure 3. Photovoltaic device fabricated by electrophoretic deposition of CdSe nanocrystals.

The light and dark I-V curves are shown in Figure 4a for the photovoltaic cells. For comparison, light and dark curves were also measured on the same cell, but in a region in which nanocrystals had not been deposited (Figure 4b). Concentrating first on Figure 4a, the cells clearly show a photovoltaic response under illumination. The open circuit voltage (V_{oc}), short circuit current (I_{sc}), fill factor (FF) and maximum power (P_{max}) for this cell were found to be: $V_{oc} = 0.36$ V, $I_{sc} = 13$ nA, $FF = 0.26$ and $P_{max} = 1.2$ nW. The response of the cell under illumination, but when nanocrystals were not present, was ohmic (Figure 4b), and shows that the photovoltaic response was due to the nanocrystals.

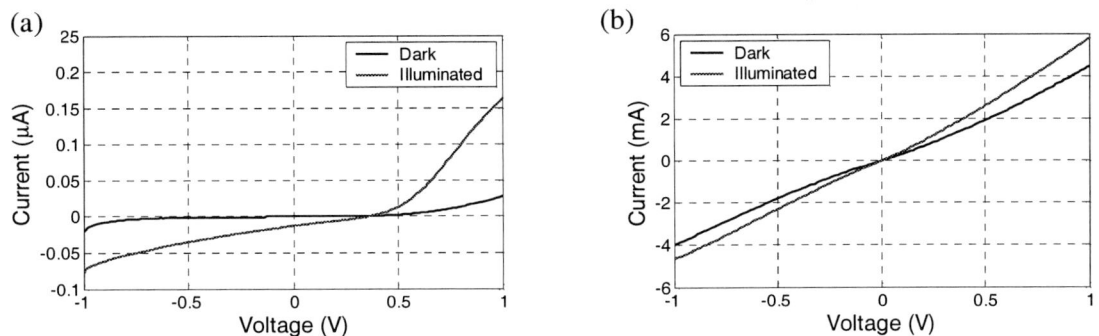

Figure 4. (a) I-V curves for a CdSe nanocrystal based photovoltaic cell fabricated by electrophoretic deposition of the nanocrystals onto a TiO$_2$ substrate. The device clearly exhibits a photovoltaic response under illumination. (b) I-V curves for the same cells, collected from a region in which nanocrystals were not deposited.

The low power developed by our cells is due to the very thin films of CdSe nanocrystals we have been able to deposit so far. The performance of the cells is likely to improve significantly with thicker films. As noted in the previous section, the deposition of nanocrystals is limited by the presence of impurities in the solution left over from the nanocrystal synthesis. We are currently exploring different synthetic techniques in order to get around this problem. We note that Islam et. al. made their nanocrystals using the dimethyl cadmium synthesis [14, 16], which gives much cleaner nanocrystals. They were able to form films greater than 0.8 μm thick using these nanocrystals (our films were too thin to measure using simple techniques such as a profilometer, and at the time of writing we have not yet prepared samples for more sensitive techniques such as AFM); we are thus confident that we will be able to increase the thickness of our films. Additionally, using a porous TiO$_2$ substrate rather than the planar TiO$_2$ films used here should also increase the efficiency of our devices.

CONCLUSION

We have shown that electrophoretic deposition is a viable method for depositing thin films of CdSe nanocrystals onto TiO$_2$ for photovoltaic applications. The photovoltaic devices presented here had low efficiencies primarily because of the small number of nanocrystals we could deposit, which in turn was caused by our inability to remove all the synthesis precursors from the nanocrystal solution. By changing synthetic methods, we believe this problem can be overcome. Utilizing a porous titanium dioxide film should also lead to greater device efficiencies.

REFERENCES

(1)   L. E. Brus, *J. Chem. Phys.* **80**, 4403-4409 (1984)

(2)   A. J. Nozik, *Physica E*, **14**, 115-120 (2002)

(3)   R. J. Ellingson, Matthew C. Beard, J. C. Johnson, P. Yu, Olga I. Micic, A. J. Nozik, A. Shabaev and A. L. Efros, *Nano Lett.* **5**, 865-871 (2005)

(4)   R. D. Schaller, M. Sykora, J. M. Pietryga and V. I. Klimov, *Nano Lett.* **6**, 424-429 (2006)

(5)   I. Robel, V. Subramanian, M. Kuno, and P. Kamat, *J. Am. Chem. Soc.* **128**, 2385-2393 (2006)

(6)   X. Qian, D. Qin, Y. Bai, T. Li, X. Tang, E. Wang and S. Dong, *J. Solid State Electrochem* **5**, 562-567 (2001)

(7)   R. Vogel, P. Hoyer and H. Weller, *J. Phys. Chemm.* **98**, 3183-3188 (1994)

(8)   L. M. Peter, D. J. Riley, E. J. Tull and K. G. Upul Wijayantha, *Chem. Commin.* 1030 (2002)

(9)   Q. Shen, D. Arae and T. Toyoda, *J. Photochem. Photobiol. Chem. A* **164**, 75-80, (2004)

(10)   O. O. Van der Beist and L. J. Vandeperre, *Annu.Rev. Mater. Sci.*, **29**, 327-352 (1999)

(11)   A. R. Boccaccini, P. Karapappas, J. M. Marijuan and C. Kaya, *J. Mater. Sci.* **39**, 851-859 (2004)

(12)   G. Kim, H. Seo, V. P. Godble, Y. Kim, O. Yang and H. Shin, *Electrochem. Commun.* **8**, 961-966 (2006)

(13)   S. V. Mahajan, D. W. Kavich, M. L. Redigilo and J. H. Dickerson, *J. Mater. Sci.* **41**, 8160-8165 (2006)

(14)   M. A. Islam and I. P. Herman, *Appl. Phys. Lett.* **80**, 3823-3825 (2002)

(15)   M. A. Islam, Y. Xia, M. L. Steigerwald, M. Yin, Z. Liu, S. O'Brien, R. Levicky, and I. P. Herman, *Nano Lett.* **3**, 1603-1606 (2003)

(16)   M. A. Islam, Y. Xia, D. A. Telesca Jr, M. L. Steigerwald and I. P. Herman, *Chem. Mater.* **16**, 49-54 (2004)

(17)   Q. Zhang, T. Xu, D. Butterfield, M. J. Misner, D. Y. Ryu, T. Emrick and T. P. Russell, *Nano Lett.* **5**, 357-361 (2005)

(18)   S. J. Rosenthal, J. McBride, S. J. Pennycook and L. C. Feldman, *Surf. Sci. Rep.* **62**, 111-157 (2007)

(19)   C. Goh, K. M. Coakley, and M. D. McGehee, *Nano Lett.* **5**, 1545-1549 (2005)

Mater. Res. Soc. Symp. Proc. Vol. 1031 © 2008 Materials Research Society 1031-H13-28

Fabrication of Copper Indium Diselenide Nanowires

Sovannary Phok[1,2], Suresh Rajaputra[1,2], and Vijay Singh[1,2]

[1]Department of Electrical & Computer Engineering, University of Kentuky, 453 Anderson Hall, Lexington, KY, 40506-0046
[2]Center for Nanoscale Science & Engineering, University of Kentucky, Lexington, KY, 40506

ABSTRACT

We report on the fabrication of copper indium diselinide nanowires (CIS NWs), using a non vacuum and low cost template-assisted technique. Highly ordered nanoporous alumina templates (AAO), a few microns thick were prepared by two-step anodization in either 0.3 M oxalic acid or 3 % sulfuric acid at room temperature. Standardized 2D AAO templates with average length of 2.5 μm and pore size ranging from 8 nm to 40 nm were fabricated. A combination of voltage ramping and pore opening in phosphoric acid was applied to remove the oxide barrier layer at the bottom of the pore in order to expose aluminum, which was used as conducting substrate for electrodeposition. CIS was deposited into the nanoporous AAO by pulse cathodic electodeposition from an aqueous acidic mixture. The electrodeposited nanowires had diameter ranging up to 40 nm depending on the dimension of the template. The nanostructured CIS/AAO were annealed in various atmospheres: (i) in vacuum at 230ºC for several hours and (ii) in flowing argon gas at temperatures ranging from 300ºC to 600ºC for a few minutes. The nanowire composition was analyzed by energy dispersive X-ray (EDX) spectroscopy. High resolution transmission electron microscopy and X-ray diffraction revealed a preferred [112] orientation.

INTRODUCTION

Nanostructured materials, including films [1,2] and nanowires (NWs) [3,4] have received a great deal of attention due to their unusual physical properties. The possibility of controlling these properties by varying the particle size, shape and surface properties is of great interest for nanoscale device applications in microelectronics, non-linear optics and optoelectronics [5] in particular. Among semiconductors of the I-III-VI group, copper indium diselenide (CIS), a p-type semiconductor with direct band gap of about 1 eV has shown promise as an absorber for photovoltaic cells. In particular, polycrystalline gallium doped CIS based solar cells have been reported to exhibit efficiency over 19% [6]. It is generally known that to improve the efficiency of a cell, an absorber layer with an optimal band gap energy tailored to solar spectrum (about 1.45 eV) should be used. One method to tailor the bandgap is by doping the semiconductor. Another method is by size tuning the CIS bandgap, where the small diameter of CIS would lead to quantum confinement. Over the past couple of decades, the fabrication of CIS films by various physical and chemical techniques has been largely explored with various degree of success. On the contrary, the synthesis of nanostructured CIS is still unexplored due to the intrinsic property of CIS to growth into grains larger than 20 nm in size [7]. Yet, nanowires with diameter smaller than 20 nm are needed to fully exploit the advantages of quantum confinement effects.

In this work, we demonstrate the feasibility of synthesizing vertically aligned CIS NWs by a simple template assisted method. Also, the technique offers the potential to control the composition and the dimension of CIS NW arrays for multifunctional nanoelectronic devices.

EXPERIMENTAL DETAILS

Preparation of AAO template by anodization

The starting material was a commercial aluminum tape (1 inch x 1 inch). The tape was cleaned in ethanol followed by de-ionized water. The anodization was performed in a solution containing either 0.3 M oxalic acid or 3% sulfuric acid in de-ionized water at room temperature. The cell is constituted of two vertical electrodes with platinum as a counter electrode. The first anodization step was performed at a constant voltage of 20 V in oxalic acid and 10 V in sulfuric acid respectively for 10 minutes. The oxide layer was removed in a hot mixture of phosphoric acid and chromic acid for several seconds. The second anodization step was performed at a constant voltage ranging from 25V to 40V in oxalic acid and at 12V-15V in sulfuric acid for less than 30 minutes. The oxide barrier layer was removed by ramping down the voltage at 1V per minute (from 10V to 1V) followed by immersion in 50% phosphoric acid at room temperature. A thermal treatment at 220 °C in air for several hours was carried out in order to remove unwanted residuals like hydroxides. Several amorphous AAO templates with pore size ranging from 12 to 40 nm were prepared by the above two step anodization.

Synthesis of CIS by pulse electrodeposition

Details of the process were recently published [8]. Briefly, the electrolyte was a mixture of copper sulfate hydrate, indium sulfate hydrate, seleneous acid and lithium chloride, dissolved in 100 mL aqueous buffer solution containing potassium hydrogen phthalate and hydrochloric acid. Electrodeposition was carried out in a two electrode cell with platinum wire as counter electrode. A first cathodic deposition was performed at 1A for few seconds to perforate any residual barrier layer. Based on previous results obtain in our laboratory, a voltage pulse of 0.7V to 1V with width of 0.3 sec. and a period of 3.3 sec. were investigated. The electrodeposited $CuInSe_2$ nanowires were annealed at 220ºC in vacuum (10^{-2} bar) to improve stoichiometry and crystallinity.

DISCUSSION

Fabrication of AAO templates

To obtain highly ordered 2D nanoporous template, we have carried out a two-step anodization in oxalic acid and sulfuric acid at room temperature. In general, the anodization process of a metal, such as aluminum, involves the (i) formation of an oxide/hydroxide layer and (ii) the dissolution of the oxide layer by the acidic electrolyte. In the case of short templates of less than 2.5μm long anodized for a short time, we found that a smaller initial "pore dimple" nanoprint from the first step required less "dissolution time" in the second step for the pore opening. Thus applying a lower voltage, i.e. 20V in oxalic acid and 10V in sulfuric acid, in the first step anodization yielded a better surface quality (fig. 1) than applying identical voltage in both steps [8]. Figure

1a and 1b show the typical top view and cross-sectional view of a 1. 5 μm AAO template anodized at 12 V in 3 % sulfuric acid and at room temperature. Figure 1c (top view) and 1d (cross-sectional view) are SEM pictures of a 2.5 μm AAO template anodized at 30V in 0.3 M oxalic acid. Obvious difference in pore diameter and pore density was observed. The typical features of the AAO templates we have produced are summarized in Table 1.

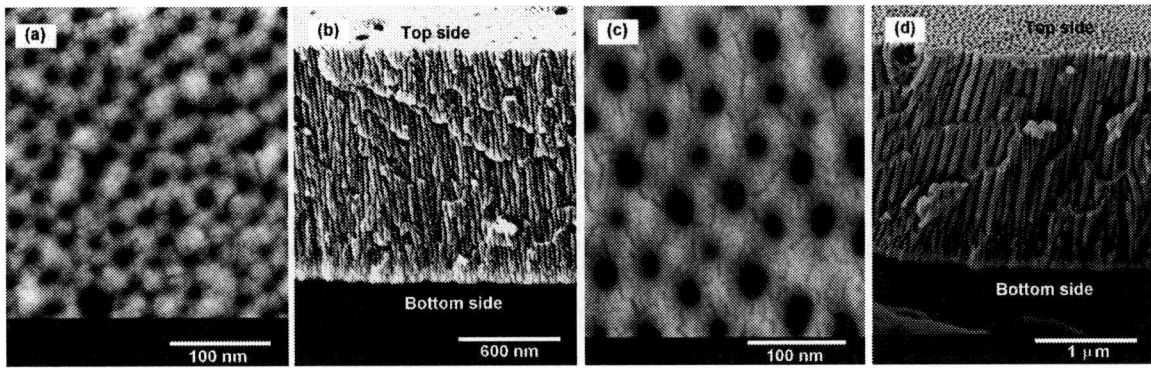

Figure 1. Typical SEM pictures of short AAO templates (< 2.5μm) anodized at room temperature in sulfuric acid at 12V; (a) top view, (b) cross sectional view and in oxalic acid at 30 V; (c) top view, (d) cross sectional view.

Electrolyte	Anodization voltage (V)		Pore diameter (nm)			Pore Density (/cm^2)
	1st Step	2nd Step	average	smaller	larger	
Sulfuric acid	10	12	12	8	30	$9 \times 10^{10} (\pm 1)$
		15	15	10	30	
		25	20	10	30	
Oxalic acid	20	30	24	15	30	$2.5 \times 10^{10} (\pm 0.3)$
		35	27	15	35	
		40	30	25	40	

Table 1. Variation of pore diameter an d pore density depending on the electrolyte and anodization voltage for a two step process at room temperature and an anodization time of less than 30 minutes.

In 0.3M oxalic acid media, the method produced templates with average pore diameter ranging from 20 nm to 30 nm with the smaller pore size at 10 nm and larger pore size at 40 nm. The approximate pore density estimated from SEM pictures was similar for the four types of template and about $2.5 \times 10^{10} (\pm 0.3)$ most likely due to an identical first step anodization at 20V. When the anodizations were carried out in 3% sulfuric acid, smaller average pore size, i.e. 12- 15 nm, were obtained. The pore density $(9 \times 10^{10} (\pm 1))$ of such templates was more than trice higher than in template anodized in oxalic acid. The smallest pore detected by SEM was about 8 nm and the largest pore was 30 nm in diameter. By adopting a step voltage ramp in order to remove the barrier layer, a root-like structure is observed at the base of the pores. High resolution SEM revealed mostly open pores at the base of a template anodized in oxalic acid. In the case of a template anodized in sulfuric acid, it was very difficult to identify the presence of a barrier layer due to small pore size.

CIS nanowires by electrodeposition

The cathodic pulse electrodeposition was performed at a voltage ranging from 0.7V to 1V with a pulsewidth of 0.3 sec and at room temperature. A strong stirring was applied during the deposition in order to avoid deposition of particles on the surface of the AAO template. The as deposited CIS NWs were annealed at 220°C for several hours at about 10^{-2} bar.

Energy dispersive X-ray (EDX) analysis was performed on a bundle of NWs embedded in AAO template. More than 2/3 of the analyzed area consisted of aluminum and aluminum oxide and this hindered an accurate analysis on CIS nanowires. However we were able to estimate the composition from the corresponding emission energy spectrum shown in figure 2. Copper, Indium and Selenium emission energies were identified. Quantitative analysis resulted in the Cu:In:Se atomic weight fractions of 20%:38%:42% whereas the atomic weight fractions of stoichiometric CIS was 19%:34%:47%. The slightly off stoichiometry was attributed to the small quantity of the materials and also due to uncertainty in the analysis of Selenium element whose emission energy is close to Aluminum. Still, within the limit of the resolution of the EDX method, the composition of these nanowires is fairly close to the stoichiometric CIS. In figure 2(c) we show a typical electron micrograph of free standing CIS nanowires after etching the AAO template.

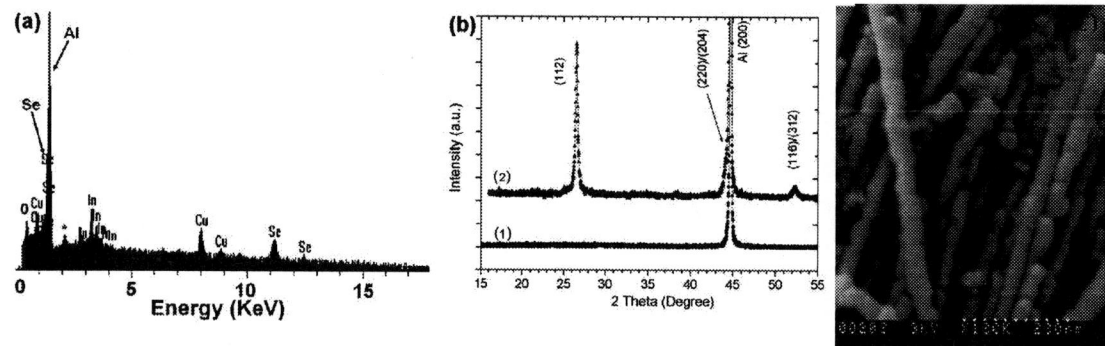

Figure 2. (a) Energy Dispersive X-Ray Spectroscopy analysis of CIS nanowires embedded in 5 µm long AAO/Al template anodized at 40V. Quantitative determination of Se element is not accurate due to peak overlapping of Al and Se. (*) indicates palladium-gold coating for SEM specimen. (b) Typical X-ray diffraction pattern of (1) AAO/Al template and CIS nanowires embedded (2) in 5 µm long AAO/Al template anodized at 40V; (c) Typical electron micrograph of a free standing CIS nanowires after etching the AAO template.

The morphology of the pulse electrodeposited CIS synthesized within an AAO template anodized at 15 V in sulfuric acid is shown in figure 3. TEM images of free standing CIS NWs showed dense, compact and unbroken NWs (fig. 3a, b, c). The average real diameter of CIS NWs fabricated within that template was about 20 nm in diameter (Figure 3b) which is slightly larger than the pore size observed by SEM. The smaller diameter of the NWs was approximately 15 nm as revealed by TEM in figure 3c.

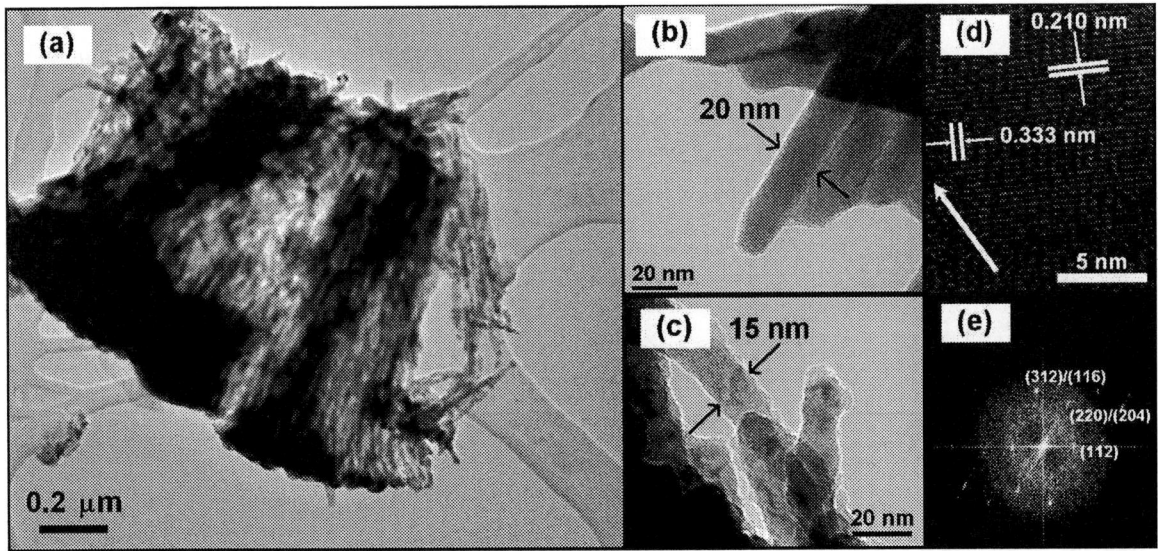

Figure 3. (a-c) TEM micrographs of free standing CIS nanowires fabricated using an AAO template anodized at 15V in sulfuric acid. Dense and compact nanowires were obtained with a length of more than 1µm. (d) HRTEM and (e) SAED pattern of a free standing CIS nanowire grown by electrodeposition in AAO template. The crystal growth direction is indicated by the white arrow.

Phases and crystal structure of the NWs were deduced from powder X-ray diffraction and HRTEM. X-ray diffraction (XRD) confirmed the presence of CIS phase. No impurity phases i.e. Cu_xSe, InSe and CuO were detected. We have indexed the XRD reflections to a chalcopyrite tetragonal phase with lattice parameters a = 5.782 Å and c = 11.619 Å according to the International Center for Diffraction Data, PDF file 00-040-1487. We have identified the main reflections at 2θ = 26.6(4)°, 44.2(0)° and 52.4(0)° corresponding to the planes (112), (220)/(204) and (312)/(116) respectively. The peak intensity indicated a preferential [112] orientation of CIS NWs. We have also performed HRTEM on a single CIS NW (Fig. 3d). The interplanar distances at 0.33(3) nm and 0.21(0) of planes perpendicular to each other were found to be related to (112) plane and (220)/(204) planes of the CIS chalcopyrite structure respectively. The corresponding SAED pattern had sharp spots with rings which was a good indication of the polycrystalline nature of CIS NWs.

CONCLUSIONS

We have synthesized nanowire arrays of I-III-VI p-type semiconductor material by a template approach. Nanoporous Alumina templates were fabricated by a two step anodization in oxalic acid and sulfuric acid. The average pore size varied from 12 nm to 30 nm depending of the applied anodic voltage. Pulsed potential electrodeposition was used to synthesize CIS inside the highly ordered nanoporous template. The simple electrochemical approach has the potential for providing a control over the characteristics of the nanowires. The technique yielded dense and compact CIS nanowires. After annealing in vacuum at 220°C for few hours, study of composition by EDX combined with structural study, i.e. XRD, HRTEM and SAED, revealed a close to stoichiometric CIS phase in a chalcopyrite crystal structure with a = 5.782 Å and c = 11.619 Å. The Cu:In:Se atomic weight fractions of 20%:38%:42% calculated by EDS along with XRD, HRTEM and SAED pattern studies were strong evidences of the existence of $CuInSe_2$ NWs grown by electrodeposition in AAO template.

ACKNOWLEDGMENTS

This work was supported in part by grants from National Science Foundation (NSF-NIRT- ECS-0609064), Army Research Laboratory, Advanced Carbon Nanotube Program (W911NF-04-2-0023) and Kentucky Science and Engineering Foundation (KSEF-148-502-03-68).

REFERENCES

1. R. S. Singh, S. Sanagapalli, V. Jayaraman, V. P. Singh, *J. Nanosci. Nanotech.* **4**, 176 (2004).
2. R. S. Singh, V. K. Rangari, S. Sanagapalli, V. Jayaraman, S. Mahendra, V. P. Singh, *Sol. Energy Mat. Sol Cells* **82**, 315 (2004).
3. A. Aguilera, V. Jayaraman, S. Sanagapalli, R. S. Singh, Vis. Jayaraman, K. Sampson, V. P. Singh, *Sol. Energy Mat. Sol Cells* **90**, 713 (2006).
4. A. K. Srivastava, R. S. Singh, K. E. Sampson, V. P. Singh, R. V. Ramanujan, *Metal. Mat. Trans. A* **38A**, 717 (2007).
5. Y. Xia, P. Yang, Y. Wu, B. Mayers, B. Gates, Y. Yin, F. Kim, H. Yan, *Adv. Mater.* **15**, 353 (2003).
6. K. Ramanathan, M. A. Contreras, C. L. Perkins, S. Asher, F. S. Hasoon, J. Keane, D. Young, M. Romero, W. Metzger, R. Noufi, J. Ward, A. Duda, *Prog. Photovolt: Res. Appl.* **11**, 225 (2003).
7. J. Muller, J. N. Nawoczin, H. Schmitt, *Thin Solid Films* **496**, 30 (2006).
8. S. Phok, S. Rajaputra, V. P. Singh, *Nanotechnology* **18**, 475601 (2007).

Hierarchical Assembly of 2D Nanostructures of Relevance for Organic Solar Cell Design

Sarah O'Donnell[1], Michael Buettner[2], and Petra Reinke[3]

[1]The Mitre Corporation, McLean, VA, 22102

[2]University of Virginia, Charlottesville, VA, 22904

[3]Department of Materials Science and Engineering, University of Virginia, 385 Mc Cormick Road, Charlottesville, VA, 22904

ABSTRACT

The first step in synthesizing a model film morphology via a surface-driven hierarchical assembly process is presented. The goal of the hierarchical assembly is the control of the morphology of complex molecular layers for the investigation of fundamental processes in organic solar cells. Using a focused ion beam (FIB) with Ga^+ ions at 30 keV, the surface of highly oriented pyrolitic graphite (HOPG) is patterned with an array of local amorphous carbon ellipsoid spots (ACES), which provide preferential nucleation lines at their perimeter, and thus are instrumental in the control of fullerene island growth. On the undamaged surface regions outside the ACES pattern the fullerene island growth is unperturbed, and presents the well-known combination of round and fractal island shapes. The fullerene deposition at the periphery of the ACES pattern, which is characterized by single ion impact defects, results in stunted, smaller and irregular islands. Inside the ACES array, the C_{60} island growth is controlled by the shape of the ACES and is constrained to lobes which form around each ACES spot. The array and C_{60} lobe morphology and geometry are characterized and a subsequent understanding of the C_{60} diffusion fields and nucleation lines within the array is discussed.

INTRODUCTION

Considerable advances in organic solar cells come from improvements in solar cell architecture and materials development[1-3]. Film morphology of the organic layers across length scales, spanning the range from a hundred nanometers down to the local molecule arrangement, has been recognized as one of the most influential material properties with respect to solar cell performance[3, 4,5,6-9]. The goal of this work is the two-dimensional morphology control of material phases in a solar cell, allowing the simultaneous investigation of fundamental aspects of exciton transport and charge diffusion which are critical components in the progression of organic solar cell construction.

Here, we use hierarchical assembly of material phases, which utilizes surface-driven routes for the synthesis of a large range of film morphologies. In this work, the first step in the hierarchical assembly for one of two material phases in an organic 2-D film is demonstrated. Using a focused ion beam (FIB), a rectangular pattern of amorphous carbon regions is created on highly-oriented pyrolitic graphite[10] (HOPG) for the self assembly of C_{60} regions to achieve control of C_{60} island morphology. This case serves as a model for testing the principles of hierarchical assembly of organic materials with control over film morphology, and several approaches to add the second molecular phase are currently being investigated. Use of a more technically relevant surface/structure, such as Y-quartz for electronic devices, will be included in subsequent experiments[11] and will pave the road towards miniature organic solar cells.

EXPERIMENT

The HOPG surface is prepared by cleaving the topmost layers using scotch tape, and adhering the sample to the slide with silver paint and spot-welded tantalum anchors. Using a focused ion beam (FIB), an array of amorphous carbon ellipsoid spots (ACES) separated by 330-350 nm is created over an area of 1 mm^2. The 30 keV Ga$^+$ ion beam is set to a current of 2 pA to mitigate sputtering, and an ion fluence of 1.5×10^{-5} ions cm^{-2}. The resulting array is shown in Figure 1b.

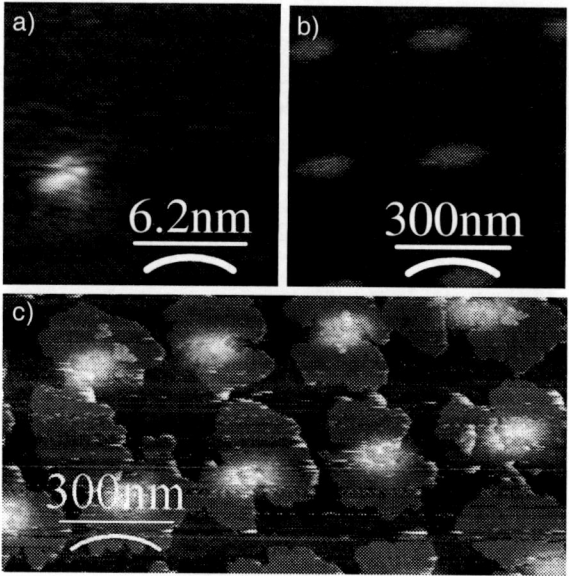

Figure 1. STM images of a) Single Ga$^+$ ion impact on HOPG, b) Ga$^+$ irradiated amorphous carbon spots (ACES) on HOPG, and c) nucleation of fullerene islands after 60 s fullerene deposition.

After preparing the pattern, the sample is annealed under vacuum using a resistive heater. The surface pattern and morphology is characterized using an Omicron Nanotechnology Variable Temperature Scanning Probe Microscope (VT-SPM) under ultra-high vacuum (UHV) conditions, 2×10^{-10} mbar base pressure. A bias voltage of 1.8 V, feedback current of 0.17 nA, and gain of 2.3% were used to produce the image in Figure 1a.

The fullerene deposition occurs in situ, with material sublimation from a fullerene source. The source is a simple thermal evaporator (tungsten wire basket with crucible). In order to image the fullerenes, a bias voltage of 1.8 V or more must be used to tunnel into the fullerene electronic states. All images of C_{60} islands, such as that of Figure 1c, is produced by using a bias voltage of 2.093 V, feedback current of 0.279 nA, and a gain of 2%.

RESULTS AND DISCUSSION

The Ga^+ ion implantation amorphizes the graphite and leads to a swelling and formation of hillocks within the ACES due to the lowered density in the amorphous region. The ACES exhibit peak heights of 5 nm on average, occurring at areas of maximum defect density. The elliptical shape of the ACES (80 nm wide and 150 nm long on average) is attributed to beam distortion due to incident angle and astigmatism, and the spread of the ion collision cascade. The angle of the ion beam incident on the HOPG surface increases with distance from the array origin. ACES close to the beam origin have aspect ratios close to one, whereas ACES at the very edge of the array area are laterally continuous with a relatively high aspect ratio, forming amorphous carbon line of periodic defect density, which can be seen in Figure 2a. Additionally the HOPG surface between the ACES includes small single ion impact spots, which exist due to incomplete blanking of the ion beam between exposures, and example for such a defect is shown in Figure 1a.

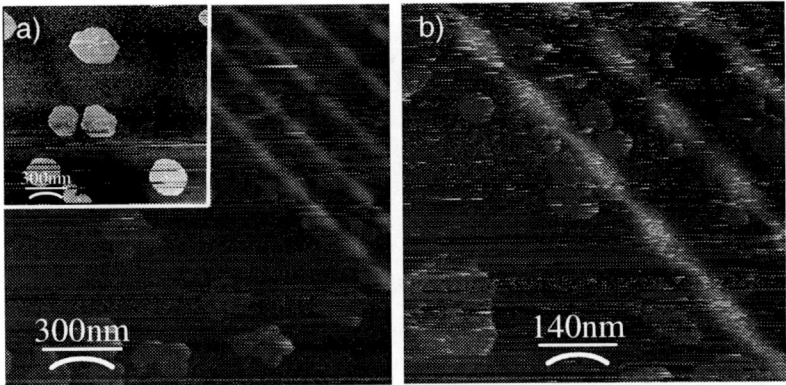

Figure 2. In a) ACES distortion creates a virtual line defect. C_{60} islands exhibit irregular island growth in the single-impact diffusion field (lower left) as opposed to unconstrained growth in an open diffusion field illustrated in the inset. b) The virtual line defect formed by the ACES limits and localizes the C_{60} island formation, illustrating an artificially constrained diffusion field.

C_{60} is deposited at a flux of 6×10^{-11} $cm^{-2}s^{-1}$ for 60 seconds resulting in sub-monolayer C_{60} coverage of 0.6 ML. The inset of Figure 2a shows island growth on HOPG, which agrees well with previous observations[12]. The island size is in this case not limited by the transport through the ACES pattern permitting islands to grow to diameters above 300 nm. The second layer island growth with the typical fractal structure is just beginning, and can be discerned on the larger islands (not shown here). C_{60} deposition on the periphery of the ACES pattern, where single ion impact defects are present, results in stunted, irregular island geometries shown on the lower-left of Figures 2a,b. The defects likely serve as nucleation centers, and at the same time modify the molecule mobility and considerably influence the diffusion field around each defect and fullerene island. Island diameters range from 150-250 nm, and their irregular shapes are an expression of the inhomogeneity in the diffusion field from random single-ion defects.

At the ACES array border, the virtual line defect created by the line of ACES produces C_{60} islands whose spatial arrangement is similar to that observed at graphite step edges[12]. Inside the array, C_{60} island growth is constrained by periodic nucleation areas provided by the ACES spot boundaries. Figure 3 shows the array of fullerene islands within the ACES array, and the Voronoi cells, which can be assigned to each island, are indicated in the figure.

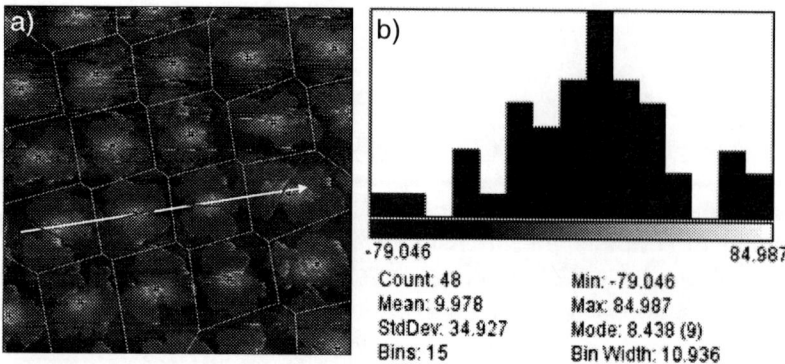

Figure 3. Pre-deposition ACES pattern orientation (white) exhibits a relative pattern angle of 3° to the horizontal. In a) post-deposition C_{60} island "lobes" exhibit a characteristic orientation (blue) with a relative angle of 1.8° to the ACES. Using Voronoi cells (yellow) on Figure 1c, C_{60} coverage is 60% monolayer. Histogram in b) shows the angle distribution of the lobes is 9.9 degrees with respect to the image horizontal, and the relative angle between ACES and lobes is 1.8 degrees.

Each ACES fullerene island contains approximately 40,000 fullerenes. Interestingly the C_{60} islands are not round but show a kidney bean-like lobe shape with the long axis perpendicular to the direction of long axis of the ACES. The ACES pattern is characterized in Figure 3a to have a mean spot angle of 9.9° with respect to the horizontal axis of the image. The orientation of the fullerene islands is then characterized by the orientation of the short axis, where the lobes meet and the island width is smallest, which is indicated by the short arrows introduced in Figure 3a. The preferred island orientation is roughly parallel to the long axis of the ACES spots. The distribution of lobe angles relative to the ACES pattern angle is presented in Figure 3b. The STM images in Figures 2 and 3 illustrate the ability of the ACES pattern to control the fullerene island growth.

The diffusion barrier[13] of C_{60} on HOPG is approximately ~13meV, and is determined primarily by van der Waals forces between the fullerene molecule and the graphite surface. Consequently the fullerene mobility is quite high on the HOPG surface. On the other hand, the ion irradiation of HOPG creates dangling bonds in the localized amorphous regions, which drastically limit the fullerene mobility. This abrupt change in molecule mobility leads to an accumulation of fullerenes at the periphery of the ACES and thus the perimeter of the amorphous spots serves as nucleation line for the growth of fullerene regions (see next paragraph). If the mean free path of the fullerenes exceeds the distance between ACES related nucleation lines, secondary nucleation will be suppressed and the islands will always start growing at the nucleation lines defined by the ACES spots. When the spots are spaced 300- 400 nm apart, as shown in Figure 2, the ACES define the nucleation of fullerene islands.

The orientation of the island is controlled by the overall pattern geometry, more precisely the edge-to-edge distance of the damage spots, which defines the width of the graphite region and thus controls the flow of fullerene molecules and the diffusion field around the island. The structured diffusion field promotes C_{60} island lobe formation, with the short axis of the lobes centered about the ACES long axis. A quantitative analysis of the diffusion field is in progress.

The transition from a high defect region at the periphery of each ACES spot to the relatively undamaged HOPG surface is shown in Figure 4. From the STM images it appears that the fullerene nucleation begins along a nucleation line, which lies in an area of transition in defect density, where there is a sharp rise in both defect density and topographical elevation. The topographical elevation is directly related to the degree of amorphization and therefore indicative of the materials defect density. This transition is shown in Figure 4. At this point, the mobility if the fullerene is kinetically limited by the increase in dangling bonds available in the amorphous region, which stabilizes the formation of a nucleus and then leads to island growth.

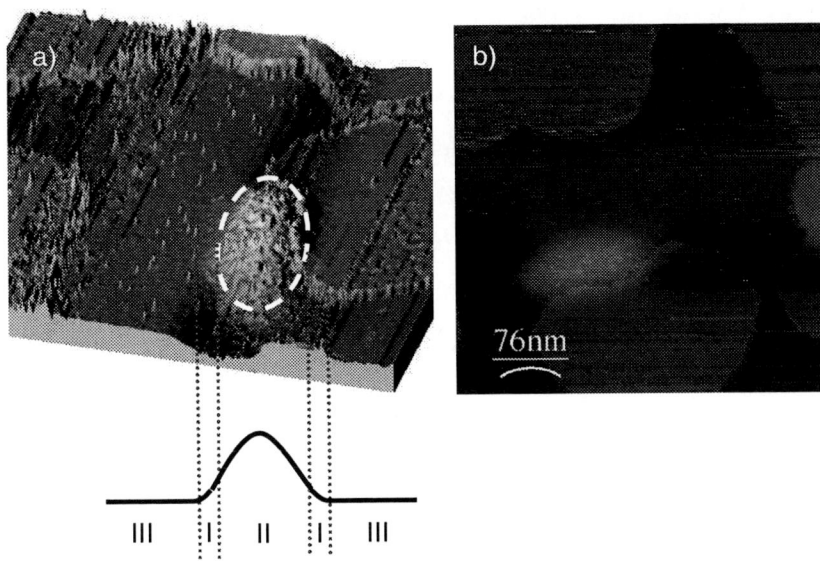

Figure 4. a) Nucleation line, indicated by dotted white line, illustrates C_{60} nucleation by the long axis of the ACES. Field I is a region of moderate defect density with a sharp slope to field II, the high defect density region. Diffusion field extends from region II to III. b) 2D representation of A. Region II spans ~5-7nm.

CONCLUSIONS

The experiments show that the first step in the hierarchical assembly of material phases is feasible and we can achieve control of fullerene film morphology and control of structure over multiple length scales. The artificial pattern enables us to design nucleation lines, which are used to control the fullerene island growth and tailor the diffusion field. A quantitative analysis of the island position and shape with respect to the pattern yields the requisite data base to tailor the fullerene island phase morphology. The next step in the hierarchical assembly will be the introduction of the second molecule phase, which will be a photoabsorber such as porphyrin. This can be achieved for example by diffusion of molecules into the fullerene pattern from a reservoir created outside the ACES region. In the present experiment we use an artificial patterning on the model surface HOPG, to test the feasibility of our approach. However, the utilization of naturally ordered structures, such as terraces of Y-cut quartz, provides a unique option for the assembly of high aspect ratio fullerene islands and the extension of the hierarchical assembly to technically relevant substrates.

ACKNOWLEDGMENTS

This work is made possible by the support of the MITRE Technology Program and the MITRE Accelerated Graduate Degree Program (AGDP). The work was supported by the MRSEC Center for Nanoscopic Materials Design sponsored by the National Science Foundation and the start-up funding from the University of Virginia (P.R.). M.B. gratefully acknowledges the financial support by the Swiss National Science Foundation.

REFERENCES

1. C. J. Brabec, N. S. Sariciftci, and J. C. Hummelen, Adv. Funct. Mater. **11**, 15 (2001).

2. J. Nelson, Current Opinion in Solid State and Materials Science **6**, 87 (2002).

3. P. Peumans, S. Uchida, and S. R. Forrest, Nature **425**, 158 (2003).

4. G. Dennler and N. S. Sacriciftci, Proc. of the IEEE **93**, 1429 (2005).

5. D. G. d. Oteyza, T. N. Krauss, E. Barrena, H. Dosch, and J. O. Osso, Appl. Phys. Lett. **90**, 243104 (2007).

6. F. Yang, M. Shtein, and S. R. Forrest, Nature Materials **4**, 37 (2005).

7. D. M. Guldi, Chem. Soc. Rev. **31**, 22 (2002).

8. H. Imahori, A. Fujimoto, S. Kang, H. Hotta, K. Yoshida, T. Umeyama, Y. Matano, and S. Isoda, Adv. Mater. **17**, 1727 (2005).

9. H. Imahori and S.Fukuzumi, Adv. Funct. Mater. **14**, 525 (2004).

10. J. Gierak, D. Mailly, P. Hawkes, R. Jede, L. Bruchhaus, L. Bardotti, B. Prevel, P. Melinin, A. Perez, R. Hyndman, J.-P. Jamet, J. Ferre, A. Mougin, C. Chappert, V. Mathet, P. Warin, J. Chapman, Appl. Phys. A **80**, 197 (2005).

11. C. Kocabas, S.-H. Hur, A. Gaur, M. A. Meitl, M. Shim, and J. A. Rogers, Small **1**, 1110 (2005).

12. H. Liu and P. Reinke, Surface Science **601**, 3149 (2007).

13. P.A. Gravil, M. Devel, Ph. Lambin, X. Bouju, Ch. Girard, A. A. Lucas, Phys. Rev. B **53**, 1622-1629 (1996).

DISCLAIMER AND LIMITER STATEMENT

The contents of this material reflect the views of the author and The MITRE Corporation and do not necessarily reflect the views of the Federal Aviation Administration (FAA) or the Department of Transportation (DOT). Neither the FAA nor the DOT makes any warranty or guarantee, expressed or implied, concerning the content or accuracy of these views.

Approved for Public Release; Distribution Unlimited.

Doped Silicon Nanoparticles Synthesized by Nonthermal Plasma

Xiaodong Pi[1], Rebecca Anthony[1], Stephen Campbell[2], and Uwe Kortshagen[1]

[1]Department of Mechanical Engineering, University of Minnesota, 111 Church Street SE, Minneapolis, MN, 55455

[2]Department of Electrical Engineering and Computer Science, University of Minnesota, 200 Union Street, Minneapolis, MN, 55455

ABSTRACT

Silicon nanoparticles (Si-NPs) < 6 nm have been doped with B and P in nonthermal plasma. The doping efficiency of B is smaller than that of P, consistent with the theoretically predicted larger formation energy of B than P. The effect of doping on the oxidation-induced changes in light emission from Si-NPs is different between B and P. It is suggested that P is at or near the surface of Si-NPs, and that B is well incorporated inside Si-NPs. Inks based on doped Si-NPs are produced by attaching alkyl ligands to the surface of Si-NPs and dispersing them in organic solvents.

INTRODUCTION

Accurately doping bulk Si to tune its properties is one of the most important achievements that have led to the success of the microelectronic industry. This has stimulated similar efforts to dope Si nanoparticles (Si-NPs). Fujii et al. [1–5] doped Si-NPs embedded in silicon oxide with B or P. They showed that the photoluminescence (PL) of Si-NPs is attenuated by B doping or high-concentration P doping. The decrease in PL efficiency resulted from the Auger effect, which involves a photogenerated exciton (electron-hole pair) and a dopant-induced free carrier (electron or hole). When P was doped at a low concentration, the PL of Si-NPs was usually enhanced. This implied that P passivated defects such as dangling bonds at the Si-NP/oxide interface. Švrček et al. [6] and Tchebotareva et al. [7] studied the effect of P doping on the PL of Si-NPs, which were also embedded in silicon oxide. Their results were similar to Fujii et al.'s.

For applications such as photovoltaics which employ the modified electrical properties of doped Si-NPs, embedded Si-NPs in a dielectric such as silicon oxide are likely not a good choice, as the matrix prohibits efficient charge transport among Si-NPs. This problem can be solved by doping free-standing Si-NPs and then making films of them. Baldwin et al. [8] have doped free-standing Si-NPs synthesized in a liquid phase with P. A P concentration of 6% in Si-NPs was achieved. Impurities resulting from their chemical reaction, however, prevented detailed characterization of the P-doped Si-NPs. In this paper we present an investigation of the doping of B or P in plasma-synthesized free-standing Si-NPs that are free of impurities except for partial coverage of H at the surface [9]. Doped Si-NP colloids (inks) have also been prepared for the formation of Si-NP films.

EXPERIMENT

The nonthermal plasma system used to synthesize intrinsic or doped Si-NPs was essentially similar to those described in previous papers [10–12]. SiH_4/Ar plasma was employed to synthesize intrinsic Si-NPs. B_2H_6/H_2 (10%/90% in volume) and PH_3/H_2 (15%/85% in volume) were added to dope Si-NPs with B and P, respectively. Si-NPs were collected on a stainless steel filter placed behind the plasma.

A transmission electron microscope (FEI Tecnai T12) was employed to examine both intrinsic and doped Si-NPs that were deposited on lacy carbon grids. Element analysis was carried out by use of an inductively coupled argon plasma optical emission (generally called ICP) spectrometer (Perkin-Elmer Optima 3000DV). The ICP characterizations of B, P and Si have all been calibrated by using standard samples. A Spex Fluorolog-2 spectrofluorometer was used to measure PL from Si-NPs that were excited at 325 nm by a Xe lamp. Si-NPs on stainless steel filters collected for a period of 10 min were used for PL measurements. The PL measurements were performed within ~ 5 min after Si-NP production and again after 5-day exposure in air at room temperature. All the PL spectra presented in this work were corrected with respect to the spectral system response.

Hydrosilylation was carried out to prepare Si-NP colloids (inks) [13]. In hydrosilylation Si-NPs without exposure in air were placed in a mixture of 1-dodecene and mesitylene (2 : 5 in volume). The reaction was performed at ~ 200 °C for enough time (> 3 h) in a nitrogen atmosphere. Hydrosilylated Si-NPs were then dried in the reactor and finally dispersed in an organic solvent to form inks.

DISCUSSION

Figure 1 representatively shows a low-resolution transmission electron microscopy (TEM) image of intrinsic Si-NPs. In certain area individual Si-NPs are discernable although they are

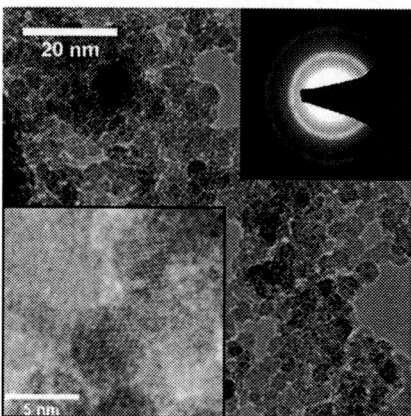

Figure 1. A low-resolution TEM image of intrinsic Si-NPs deposited on a lacy carbon TEM grid. A high-resolution TEM image is shown as the inset at the bottom-left corner. The inset at the top-right corner illustrates a selected area electron diffraction pattern.

generally agglomerated via Van de Walls force. Si crystal lattice fringes with a spacing of 0.31 nm are seen in the high-resolution TEM image, which is illustrated as one of the insets of figure

1. The other inset illustrates a selected area electron diffraction pattern, which further indicates the crystallinity of Si-NPs. A statistic analysis on the size d of Si-NPs gives $d = 3.6 \pm 0.8$ (standard deviation). We note that doping does not affect the size of Si-NPs.

Assuming a doping efficiency of 100%, we can calculate the ideal atomic concentrations C_i of dopants in Si-NPs from gas flow rates. The actual atomic concentrations C_m of dopants are obtained from ICP measurements. Figure 2 shows the doping efficiency (C_m/C_i) of (a) B or (b) P. It is clear that for P doping the doping efficiency is ~ 100% when C_i is smaller than ~ 4 %. When C_i is further increased up to 9%, the doping efficiency is reduced to ~ 63 %. This means that it is less efficient to dope Si-NPs with P at high concentrations, likely owing to the condensation of PH_3 at the tube wall [14]. It is interesting to note that for B doping the doping efficiency is > 10 times smaller than that of P. A recent computational study has demonstrated that in Si-NPs the formation energy of substitutional B is larger than that of substitutional P when structure relaxation after doping is not involved [15]. Such a prediction is consistent with our present observation that B is less efficiently doped than P.

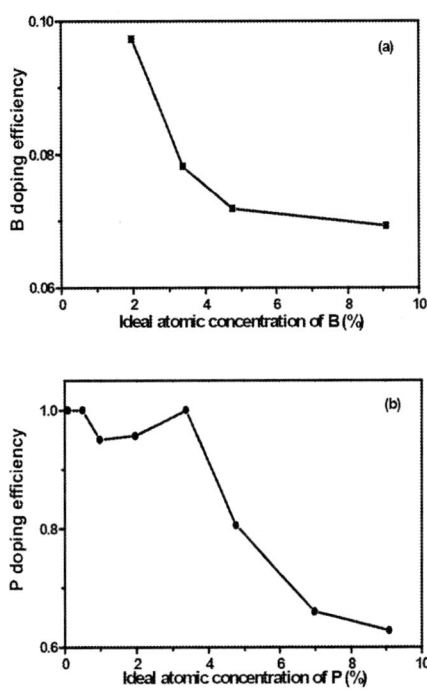

Figure 2. Doping efficiency versus ideal dopant atomic concentration for (a) B and (b) P. Solid lines are drawn to guide the eye.

Figure 3 shows the PL spectra from (a) as-synthesized intrinsic and B-doped Si-NPs and (b) the same Si-NPs exposed in air at room temperature for 5 days. B in Si-NPs may be ionized to generate free holes [1, 2]. This gives rise to Auger recombination, which can quench the light emission from excitons in Si-NPs. Therefore, the intensity of PL (I_{PL}) from Si-NPs decreases with B doping. With an increase of B concentration up to 0.34% the rate of recombination via the Auger effect increases, leading to a decrease in I_{PL} by a factor up to ~ 100. Figure 3 (b) shows that after 5 days in air the PL from both intrinsic and B-doped Si-NPs blueshifts. This is due to the fact that the size of Si-NPs is reduced after oxidation in air [9, 12, 16]. It has been suggested that Si-NP/oxide interface is less defective than the surface of as-synthesized Si-NPs [12]. This explains the increase in I_{PL} for all the Si-NPs after the oxidation. Despite the overall increase in I_{PL} we notice that B-doped Si-NPs still emit light less efficiently than intrinsic Si-NPs. We propose that B is incorporated well inside Si-NP. After the oxidation-induced conversion of the near surface region to silicon oxide, there is remaining B in the interior of Si-NPs, as evidenced by its attenuating effect on the PL.

The PL spectra from (a) as-synthesized intrinsic and P-doped Si-NPs and (b) the same Si-

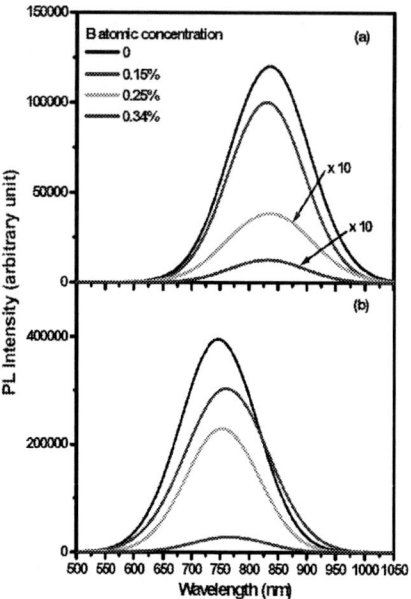

Figure 3. PL spectra from (a) as-synthesized intrinsic and B-doped Si-NPs and (b) the same Si-NPs after 5-day exposure in air at room temperature. The B-doped Si-NPs are identified by B atomic concentrations obtained from ICP measurements. The intensity of PL from Si-NPs with B concentrations of 0.25% and 0.34% is magnified by a factor of 10.

NPs exposed in air at room temperature for 5 days are shown in figure 4. When P concentration is 0.06%, The I_{PL} of P-doped Si-NPs is larger than that of intrinsic Si-NPs. This increase in I_{PL} indicates that P intends to segregate at the surface of Si-NPs, directly passivating surface defects such as dangling bonds [3, 4]. Investigations on doping Si in molecular beam epitaxy (MBE) have shown that the segregation of dopants such as B or P at the surface of Si is determined by surface diffusion at low temperatures [17–19]. The temperature of Si-NPs during their nucleation and growth is low (< 600 °C) in nonthermal plasma [10]. We believe that after the initial nucleation, Si-NPs grow in an epitaxial manner. The segregation coefficient κ, which is the ratio of dopant concentration at surface to that in bulk during epitaxy, of P is much larger than that of B [17]. This explains that P predominantly stays at the surface region of Si-NPs, in contrast to the case for B. At a given κ the number of P atoms inside Si-NPs increases with the increase of total P concentration. Beyond certain P concentration free electrons induced by the ionization of P lead to Auger recombination [4, 5]. Therefore, the I_{PL} of P-doped Si-NPs becomes smaller than that of intrinsic Si-NPs when the concentration of P is $\geq 0.4\%$. Similar to what happens in the B doping, the PL is attenuated more significantly with the further increase of P concentration up to 5.6%. As pointed out in ref. [7], the preferential doping for larger Si-NPs in a Si-NP ensemble may give rise to the blueshift of PL when the concentration of P increases in the high-concentration range. In the low-concentration range for both P and B, the effect of preferential doping is not pronounced, and hence the peak position of PL is nearly unchanged.

After 5 days in air the I_{PL} of all the P-doped Si-NPs is comparable to that of intrinsic Si-NPs, implying that P-induced surface passivation and Auger recombination have ceased as a result of oxidation. This suggests that the large κ of P does cause all the P to be located at or near surface. After the oxidation P may be encapsulated in silicon oxide and become inactive, exhibiting no effect on the PL from Si-NPs.

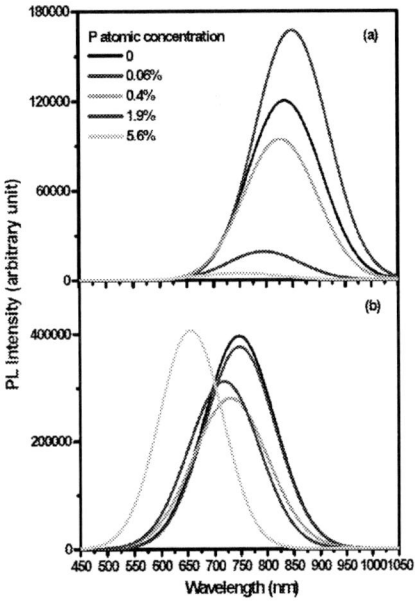

Figure 4. PL spectra from (a) as-synthesized intrinsic and P-doped Si-NPs and (b) the same Si-NPs after 5-day exposure in air at room temperature. The P-doped Si-NPs are identified by P atomic concentrations obtained from ICP measurements.

We have obtained both intrinsic Si-NP inks and B- or P-doped Si-NP inks by means of hydrosilylation. The light emission efficiency for hydrosilylated Si-NPs are significantly larger than that for unhydrosilylated ones. However, the effect of doping on the PL from Si-NP inks is similar to the above discussed results for as-synthesized intrinsic and doped Si-NPs in the form of powder collected on stainless steel filters. The alkyl ligands attached to the surface Si-NPs does not fully stop the oxidation of Si-NPs in air. The oxidation of hydrosilylated Si-NPs usually leads to the decrease in light emission efficiency. Detailed information on the mechanism and process of the hydrosilylation of Si-NPs, and the optical properties of hydrosilylated Si-NPs may be found in [13] and [20].

CONCLUSIONS

In summary, we find that the doping efficiency of B is smaller than that of P, consistent with the larger formation energy of B than P. Subsequent to the initial incorporation at certain lattice site, B appears to be less mobile by means of surface diffusion than P. Correspondingly, B is incorporated well inside Si-NPs while P is predominantly located at or near the surface of Si-NPs. Doped Si-NP inks may be obtained after hydrosilylation in a scheme similar to that for intrinsic Si-NP inks.

ACKNOWLEDGMENTS

The authors are grateful to Professor David J. Norris for support in the PL measurements. Russel Anderson is thanked for the ICP measurements. This work was supported primarily by the MRSEC Program of the National Science Foundation under Award Number DMR-0212302, and partially by SPAWAR and NSF grant DMI-0556163.

REFERENCES

1. S. H. Y. Kanzawa, M. Fujii and K. Yamamoto, Solid State Commun. 100, 227 (1996).
2. M. Fujii, S. Hayashi, and K. Yamamoto, J. Appl. Phys. 83, 7953 (1998).
3. S. H. M. Fujii, A. Mimura and K. Yamamoto, Appl. Phys. Lett. 75, 184 (1999).
4. A. Mimura, M. Fujii, S. Hayashi, D. Kovalev, and F. Koch, Phys. Rev. B 62, 12625 (2000).
5. M. Fujii, A. Mimura, and S. Hayashi, Phys. Rev. Lett. 89, 206805 (2002).
6. V. ˇSvˇcek, A. Slaoui, J.-C. Muller, J.-L. Rehspringer, B. Honerlage, R. Tomasiunas, and I. Pelant, Physica E 16, 420 (2003).
7. A. L. Tchebotareva, M. J. A. de Dooda, J. S. Biteen, H. A. Atwater, and A. Polman, J. Lumin. 114, 137 (2005).
8. R. K. Baldwin, J. Zou, K. A. Pettigrew, G. J. Yeagle, R. D. Britt, and S. M. Kauzlarich, Chem. Commun. 6, 658 (2006).
9. X. D. Pi, L. Mangolini, S. A. Campbell, and U. Kortshagen, Phys. Rev. B 75, 085423 (2007).
10. L. Mangolini, E. Thimsen, and U. Kortshagen, Nano Lett. 5, 655 (2005).
11. L. Mangolini and U. Kortshagen, Adv. Mater. 19, 2513 (2007).
12. X. D. Pi, R. W. Liptak, S. A. Campbell, and U. Kortshagen, Appl. Phys. Lett. 91, 083112 (2007).
13. D. Jurbergs, E. Rogojina, L. Mangolini, and U. Kortshagen, Appl. Phys. Lett. 88, 233116 (2006).
14. P. Rai-Choudhury and F. I. Salkovitz, J. Cryst. Growth 7, 361 (1970).
15. G. Cantele, E. Degoli, E. Luppi, R. Magri, D. N., G. Iadonisi, and S. Ossicini, Phys. Rev. B 72, 113303 (2005).
16. G. Ledoux, O. Guillois, D. Porterat, C. Reynaud, F. Huisken, B. Kohn, and V. Paillard, Phys. Rev. B 62, 15942 (2000).
17. J. F. Nutzel and G. Abstreiter, Phys. Rev. B 53, 13551 (1996).
18. L. Oberbeck, N. J. Curson, T. Hallam, and M. Y. Simmons, Appl. Phys. Lett. 85, 1359 (2004).
19. H. Jorke and H. Kibbel, Appl. Phys. Lett. 57, 1763 (1990).
20. F. J. Hua, M. T. Swihart & E. Ruckenstein, *Langmuir* **21**, 6054 (2005)

Mater. Res. Soc. Symp. Proc. Vol. 1031 © 2008 Materials Research Society

III-V Semiconductor Vertical and Tilted Nanowires on Silicon Using Chemical Beam Epitaxy

Gokul Radhakrishnan[1,2], Alex Freundlich[1,2], Joe Charlson[1], and Bodo Fuhrmann[3]

[1]Electrical and Computer Engineering Department, University of Houston, Houston, TX, 77204-5004
[2]Center for Advanced materials, Houston, TX, 77204-5004
[3]Interdisciplinary Center of Materials Science, Martin Luther University of Halle, Halle, D-06120, Germany

ABSTRACT

Nowadays nanostructures play a vital part in the rapidly expanding areas of photovoltaics. The ability of nanowires to transfer photo-generated carriers rapidly across a solar cell has lead to our interest in growth of nanowires. Currently Vapor Liquid Solid (VLS) epitaxy is the most common method used to grow vertically aligned nanowires. A metal particle such as gold is used to form a liquid alloy eutectic with the material of a substrate or with material supplied in the vapor phase. In growing semiconductor wires using metal droplets, it has been shown that the wires grow in the (111) direction and have clean facets. Furthermore these wires generally present an undesirable larger pyramidal base at the bottom and there is also evidence of surface migration of the metal catalyst. Thus far, most of the effort in the development of vertical III-V semiconductor nanowires has been limited to homo-polar combinations (e.g. InAs on InP). The ability to fabricate III-V nanowires on silicon could however pave the way toward the monolithic integration of III-V nanostructured solar cells with Si. Here we demonstrate the growth of GaAs and InP nanowires on silicon (111) using gold as the metal seed particle. An ordered array of gold nano dots was patterned on the surface of a silicon substrate using self-assembled polystyrene nanospheres as the Au evaporation template. The size of the gold dots range from 40 nm to 150 nm and the pitch is about 500 nm. The growth of the nanowires was performed by chemical beam epitaxy under a vapour phase environment. Scanning electron microscopy, photoluminescence and Raman spectroscopy were used to characterize these nanowires. These nanowires exhibit high crystallinity and there is an absence of the pyramidal base at the bottom of the nanowire using this technique. Furthermore the study also shows evidence of pre-growth motion of some of the gold particles causing coalescence of nanowires and leading to the development of nanopods and tilted (off-normal) nanowires. Finally in light of their optical properties the relevance of these wires to photovoltaic applications is discussed.

INTRODUCTION

The most successful approach towards nanowire generation is by using the vapor-liquid-soild (VLS) mechanism [1]. The mechanism is based on breaking the isotropic symmetry of the crystal to a so called anisotropic crytal growth. A metal particle such as gold is used to form a liquid alloy eutectic with the material of the substrate or with material supplied in the vapor phase. The supersaturated liquid alloy leads to 1D nanowire formation in one preferential direction. The liquid alloy remains at the tip of the wire acting as a catalyst leading to longer nanowires.

Nanowires have been fabricated using different growth techniques including metal-organic vapor phase epitaxy (MOCVD) [2], laser-ablation [3], molecular beam epitaxy (MBE) [4] and chemical beam epitaxy (CBE) [5]. Here we demonstrate nanowire growth using chemical beam epitaxy using a metal seed particle. The formation and distribution of these metal nanoparticles in a uniform and reproducible manner was the main challenge. Several approaches have been used to obtain metal growth sites at nanoscale, such as mask templates of porous alumina [6,7] , aerosol technique [8] electron beam lithography(EBL) [9] and nanosphere lithography (NSL) [10] or natural lithography[11,12].

Nanosphere Lithography is an attractive approach due to its potential to form uniform metal nanoparticles across the substrate through an inexpensive, parallel, bench top technique. NSL masks are made by self-assembly of polystyrene nanospheres over the surface of the substrate. The diameter of the nanospheres ranged from 100-750nm depending on the substrate and growth material.

Development of vertical III-V semiconductor nanowires thus far has been limited to homo-polar combinations (e.g. InAs on InP). The ability of the fabrication of III-V nanowires on silicon could however pave the way toward the monolithic integration of III-V optoelectronics with Si electronics. In this paper we investigate the growth of GaAs and InP nanowires on silicon (111) using gold as the metal seed particle.

EXPERIMENT

Si(111) substrates (As-doped, resistivity of <0.005 Ωcm and P-doped, resistivity of 1-20 Ωøcm, respectively) were cut in to 2 X 2cm^2 pieces and used in all experiments. In order to make the surface of the silicon hydrophilic the substrates were cleaned by RCA I process, that is, a treatment with a 1:1:5 solution of NH_4OH (25%), H_2O_2 (30%), and water at 80 °C for 15 min just before usage.

Monodisperse polystyrene particles (purchased from Microparticles (GmbH Berlin) with diameters of 488 nm (polydispersion index PDI 0.11), 600 nm (PDI 0.08), 780 nm (PDI 0.08) respectively, were received as a 10 wt % suspension in water. They were further diluted with methanol containing 0.25% Triton X-100. For spheres up to 1040 nm, the corresponding suspensions were then spin-coated onto the wafers according to the technique described by Hulteen et al [13] Depending on the particle concentration and the spin speed, monolayer and bilayers of spheres were formed. In contrast to masks of a monolayer of polystyrene spheres, where the deposited metal particles are arranged in a hexagonal two-dimensional lattice with a two-point basis (honeycomb structure), a primitive hexagonal arrangement of the metal clusters is obtained for polystyrene bilayers[14]. It should be mentioned that the periodic arrangement of the polystyrene spheres might be locally disturbed. The most serious defects are grain boundaries.

These monolayer or bilayer arrangements are used as masks over which gold is deposited by thermal evaporation. A 10 to 20nm thick layer of gold is deposited at a rate of 0.1nm/sec in an evaporation system (B 30.2, HVT Dresden) with chamber pressure of 5 X 10^{-6} mbar. The gold particles penetrate through the mask to form a nanoscale pattern on the substrate surface. The polystyrene masks are removed from the substrate by (i)

CH_2Cl_2 in an ultrasonic bath for 2 min and (ii) by a subsequent rinsing in acetone, ethanol, and water. Synthesis of masks on hydrophobic surfaces can be done by using an elegantly simple mask transfer technique [15].

The substrate with patterns of gold nanoparticles is introduced into a RIBER ultrahigh vacuum CBE chamber. TMI (trimethyl indium) and TGA (triethylgallium) were used as group III precursors. Arsine and phosphine are used as group V precursors. Arsine and phosphine are thermally cracked at 900 °C on entering the growth chamber to the highly reactive As_2 and As_4 molecules which either incorporates at the position of impact i.e. at the gold particle or desorbs immediately from the surface and is pumped away. The group III precursors (TMI, TGA) are the stable molecules that react at the heated gold particle leading to supersaturation and the formation of the eutectic alloy. When the eutectic system is created, the driving force for wire growth is the supersaturation of the growth species in the molten seed. The growth species is fed continuously to maintain supersaturation and aid nanowire growth. The substrate was heated to 540 °C during GaAs wire growth with a growth rate of 1micron/hour and 420°C for InP wires with a growth rate of 0.2 micron/hour.

RESULTS AND DISCUSSION

The substrates were characterized by using a scanning electron microscope (SEM - LEO 1525 Carlzeiss). The nanosphere lithography process creates both single layer and double layer masks depending on spin speed and nanosphere concentration. A single layer mask creates a honeycomb structure with triangular gold patterns (pyramidal) as shown in figure 1(a) with red circles indicating the position of the polystyrene spheres. Double layer masks create hexagonal structures with circular gold patterns as shown in figure 1(b).

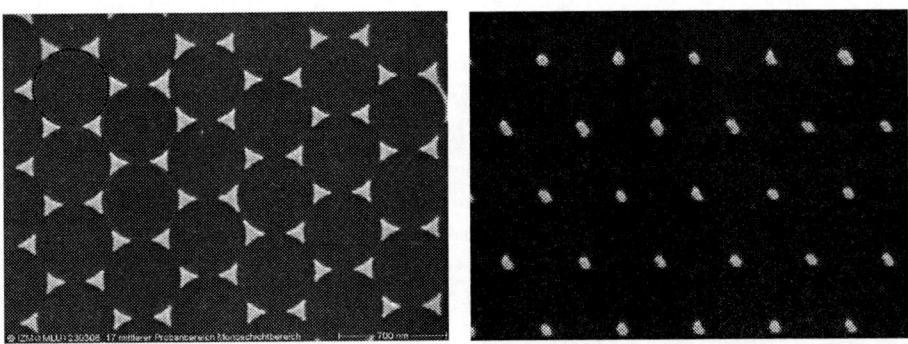

Figure 1 (a) Monolayer NSL gold pattern, (b) Bilayer NSL gold pattern

Annealing studies were performed to analyze the change in shape of the gold patterns over the Si (111) surface. The Pyramidal patterns of the gold changed in shape and size to a hemispherical globule like structure with diameter half the length of the triangular base of the pyramid. Triangular patterns of size ~120nm became hemispherical structures of diameter ~60nm after annealing in a nitrogen environment for ten minutes at 500°C as shown in figure 2(a) and (b) respectively.

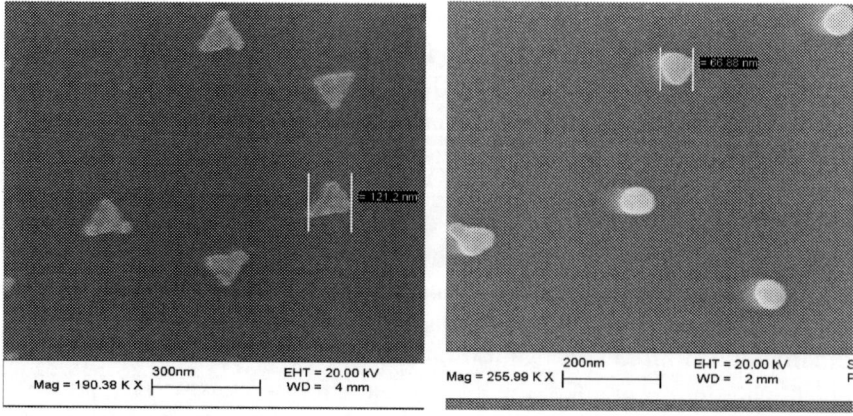

Figure 2 (a) Pyramidal structure prior to annealing (b) hemispherical structure post annealing

The substrate patterned with the gold nanoparticles was introduced into the CBE chamber. VLS mechanism of nanowire growth commences here. The substrate was heated to a temperature higher than the growth temperature with plenty of group V material (As, Ph) in vapor phase. When the group III (Ga, In) material is introduced in the chamber the gold nanoparticles act as a sinks collecting growth species or adatoms forming a liquid eutectic alloy at the vapor- liquid interface. The temperature was reduced to the growth temperature of 420^0C and 480^0C for InP and GaAs respectively. When more growth material is supplied at the growth temperature, the alloyed particle becomes supersaturated leading to crystallization at the liquid solid interface. When the temperature is reduced to the growth temperature, the liquid eutectic will try to maintain the supersaturation leading to the precipitation of the growth material at the liquid –solid interface and the wire starts to grow. Through out the growth process the gold seed particle will not be consumed and therefore will partake as a physical catalyst for 1 D growth. The gold remains at the tip of the wires at the end of growth.

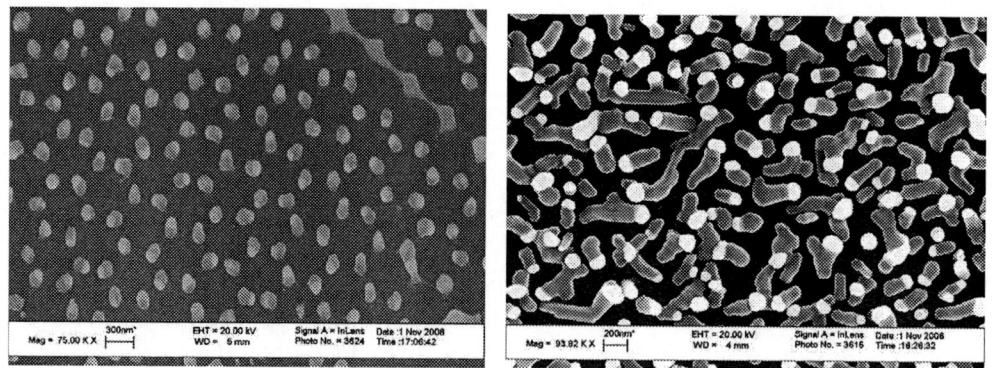

Figure 3 (a) SEM image of InP nanowires on Si (111) and (b) GaAs nanowires and nanopods on Si (111)

34

Figure 3 (a) & (b) shows the SEM images of InP nanowires on Si(111) and GaAs nanowires and nanopods on Si(111). The thin oxide layer above the Silicon substrate was not removed prior to growth. The growth rates of InP nanowires was slower than the GaAs. In the figure 3 the gold nanoparticle single layer honeycomb structure is maintained in InP wires and as the wires grow longer in case of GaAs the basic nanoparticle structure disappears. In growing semiconductor wires using metal nanoparticles it has been shown before that the wires grow in the (111) direction i.e. normal to the substrate surface and have clean facets [16, 17]. But here most of the wires tend to grow in the (100) direction. It has also been shown that these wires generally present an undesirable larger pyramidal base at the bottom [18]. Here in our SEM pictures there is absence of the pyramidal base at the bottom all of the InP and GaAs nanowires.

PL spectra of the Si(111) sample with InP wires is shown in figure 4. The PL measurements were done at 10K. The center of the PL peak is at 1.36ev. This peak is attributed to the InP wires and there is a 40mev shift from bulk InP photoluminescence peak. This shift may be due to the presence of impurities or due to residual strain on the InP wires.

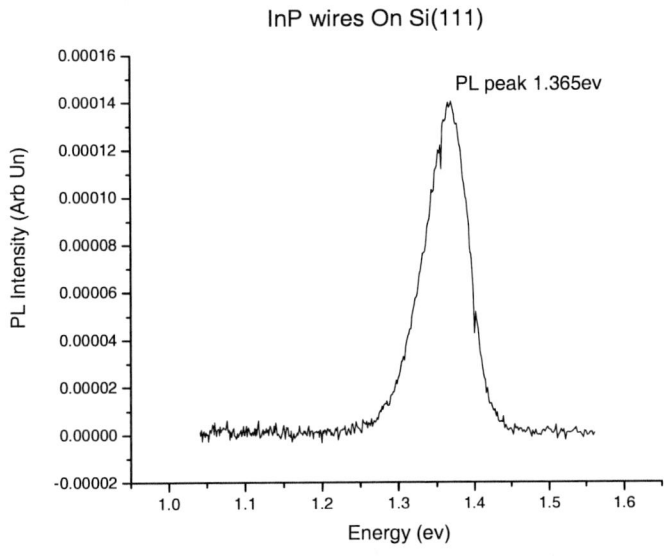

Figure 4. PL spectra of InP nanowires on Si(111) samples at 10K

Even though the gold seed particle acts as a catalyst and do not get consumed during the actual growth process, there is evidence showing migration of gold along the wire sidewalls, surface of the substrate and in the vapor phase [19]. The migration of the gold happens prior and during the growth process. Gold particles from smaller dots diffuse to larger dots leading to wires of different length shape and diameters. The gold at the tips of some wires even disappear and combine with closer wires to form large nanopods. The coalescence of the wires happens during growth resulting in a mixture of nanowires and nanopods as seen clearly in figure 3(b) of the GaAs nanopods.

CONCLUSION

Fabrication of III-V semiconductor nanowires on Si (111) was attempted. This paves way for a new approach of integrating the already existing Si electronics with III-V materials possessing immense optoelectronic properties. Nanosphere lithography has been shown to be a simple, bench top, cost effective method with excellent control of particle spacing towards the formation of growth sites for nanowires. Annealing of the gold nanoparticle leads to 50% decrease in particle size. The wires grow in (100) direction with no pyramidal base at the bottom. There is also evidence for movement of gold particle prior and during growth leading to wires of different diameter and the formation of nanopods.

REFERENCES

[1] Wagner R. S et al, Appl Phys Lett **4** 89-90 (1964)
[2] J. Johansson et al, J. Phys. Chem. B **109**, 13567–13571 (2005)
[3] Wu, Y et al, NanoLetters **2**,83, (2002)
[4] Wu Z. H et al Appl.Phys. Lett **81**, 5177, (2002)
[5] L. E. Jensen et al, NanoLetters **4**, 1961-1964 (2004)
[6] Xiao-Ling Che et al, Appl Phy Lett **88**, 263107, (2006)
[7] H. Chik et al, Appl. Phys Lett **84**, 3376, (2004)
[8] M.H. Magnusson et al, J. Nanoparticles Res **1** 243 (1999)
[9] H.T. Ng et al, NanoLetters **4** 1247 (2004)
[10] Hulteen. J. C et al, J. Vac Sci Technol. A **13**, 1553, (1995)
[11] Fischer U. C. et al, J. Vac Sci Technol **19**, 881, (1982)
[12] Deckman, H.W. et al, Appl Phys Lett **41**, 377, (1982).
[13] Hulteen, J. C et al, J. Vac. Sci. Technol., A **13**, 1553, (1995)
[14] Bodo Fuhrmann et al, Nanoletters **5** 2524 2005
[15] Hong Jin Fan et al, J of crystal growth **287**, 2006, 34-38
[16] Wagner R. S et al, Appl Phys Lett **4** 89-90 (1964)
[17] Linus E. Jensen et al, Nanoletters **4** 1961 (2004)
[18] A.I. Persson et al, Journal of Crystal Growth **272** 167-174 (2004)
[19] J. B. Hannon et al, Nature – Vol 440 69 (2006)

Mater. Res. Soc. Symp. Proc. Vol. 1031 © 2008 Materials Research Society   1031-H13-18

Analysis of Strain Compensation in Quantum Dot Embedded GaAs Solar Cells

Christopher Bailey[1], Cory Cress[1], Ryne Raffaelle[1], Seth Hubbard[1], William Maurer[2], David Wilt[2], and Sheila Bailey[2]

[1]NanoPower Research Laboratory, Department of Physics, Rochester Institute of Technology, 85 Lomb Memorial Drive, Rochester, NY, 14623

[2]NASA Glenn Research Center, Cleveland, OH, 44135

ABSTRACT

The effects of strain within stacked layers of InAs quantum dots (QDs) were investigated. InAs QD test structures with and without strain compensation (SC) were analyzed using atomic force microscopy, transmission electron microscopy, and X-ray diffraction. The affects of strain compensation on test structure morphology and on GaAs-based QD solar cell performance was studied as a function of the thickness of the SC layer. X-ray diffraction analysis of the QD embedded test structures reveals a relationship between the SC thickness and the observed crystalline quality. Air mass zero illuminated current vs. voltage data and spectral responsivity measurements were used for the solar cell comparison. When SC is employed, QD insertion shows a lower open circuit voltage, in reference to a baseline device without QDs, but leads to an enhancement in the short circuit current of the device.

INTRODUCTION

Ultra high efficiency InGaP/GaAs/Ge triple-junction solar cells (TJSC) are typically current limited by the middle GaAs junction [1]. Improving the production of photocurrent in this junction will improve the global conversion efficiency of the triple junction solar cell. There are a number of possible means to improve the current production of the middle GaAs junction including lattice-mismatched metamorphic InGaAs [2], multi-quantum wells [1] and quantum dot arrays [3]. All of these methods use lower bandgap materials inserted into the GaAs cell in order to enhance long-wavelength absorption and thus increase the short circuit current (I_{SC}). It has been predicted that quantum dot enhanced TJSCs have an efficiency ceiling of 47% under one-sun 6000 K black body illumination spectrum [4]. Additionally, quantum dot array enhanced GaAs cells have the added benefit of possible intermediate band effects [5], anisotropic absorption [6] and enhanced radiation tolerance [7].

In the case of epitaxially grown QD structures, the growth mode is usually strain driven (Stranski-Krastanow technique) [8]. Despite the benefits of the additional absorption, the build-up of strain due to QD growth has been previously shown to induce dislocations, degrading the solar cell I_{sc} and open circuit voltage (V_{oc}), and thus leading to overall device degradation [9, 10]. However, strain compensation can be used effectively to balance the residual strain, impede dislocation formation, and lead to improvements in the solar cell characteristics [11, 12]. In this paper, we further investigate the effects of strain compensation by comparing uncompensated InAs QD solar cells and strain compensated InAs QD solar cells by both transmission electron microscopy (TEM) and high resolution X-Ray diffraction (HRXRD). These techniques are used to verify the material improvements due to strain compensation and related them to device performance, specifically improvements in air mass zero (AM0) current density vs. voltage curves and spectral responsivity.

EXPERIMENT

Seven samples were epitaxially grown using a low-pressure Veeco D125LDM organometallic vapor phase epitaxy (OMVPE) system. The growth of InAs quantum dots was performed using the SK growth technique. Growth conditions and optical properties of the QDs have been reported previously [8]. In order to compensate for the InAs compressive strain (7.8% compressive mismatch), layers of tensilely strained GaP (3.2% tensile mismatch) were grown in between successive QD layers. Four samples (Q1, Q2, Q3 and Q4) were grown with InAs QD test structures and 4 samples (S1, S2, S3 and S4) were grown with the GaAs single junction *pin* solar cell structures shown in Figure 1. Reference structure (Q1) was grown with no strain compensation (SC), consisting of 5 layers of InAs quantum dots separated by 10 nm layers of undoped GaAs. Three experimental samples (Q2, Q3 and Q4) were grown with the addition of GaP inside the 10 nm GaAs capping layer. Q2 was a stack of 5 layers, and Q3 & Q4 were grown with 10. These samples were grown with GaP strain compensation layer thicknesses of 14 Å (Q2 and Q3) and 18 Å (Q4). TEM and AFM images were then taken of all four samples and results analyzed and compared.

Solar cell device S1 was grown with 5 layers of InAs QDs but without SC, while the solar cell samples S2 and S3 contained 5 layers of InAs QDs embedded in the i-region (see Figure 1). S4 was our standard baseline solar cell device structure without quantum dots. Lattice-matched InGaP window layers were included below the base and above the emitter to reduce back/front interface recombination. The i-region of the quantum dot cell included an undoped five-stack array of SC and non-SC InAs quantum dot/GaAs clad layers, totaling a thickness of 100 nm. Growth conditions for embedded dots were identical to respective test structures.

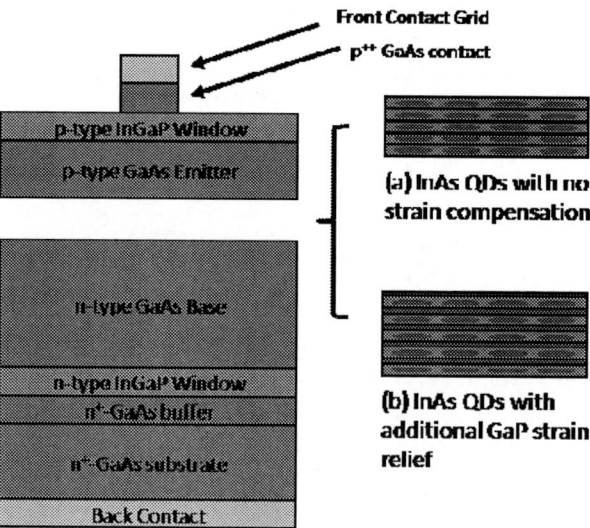

Figure 1. Solar cell layers showing quantum dot layers grown between the base and emitter without (a) or with (b) the SC GaP layer between successive dot arrays inserted.

Solar cells were fabricated using standard semiconductor processing and lithography techniques. Au/Ge and Au/Zn metallizations were used for p-type and n-type ohmic contacts, respectively. The 1 x 1 cm^2 solar cells fabricated for this experiment had total grid finger area of 0.04 cm^2. Antireflective coatings were not used for this study.

Solar cell current-voltage data was measured under one-sun air mass zero (AM0) illumination conditions provided by a 1000W Oriel solar simulator. Verification of one sun AM0 illumination is performed before all testing, using a calibration cell supplied by the NASA Glenn Research Center. Solar cell spectral responsivity measurements are obtained by a high-intensity source attached OL Series 750 Spectroradiometric Measurement System. Reflectance measurements were taken using a Perkin Elmer Lambda 900 UV/VIS/NIR spectrometer with an optical integration sphere. High resolution X-ray diffraction spectra were obtained using a Bruker AXS D8 Discovery system.

DISCUSSION

Quantum Dot Characterization

Transmission electron microscopy was employed to reveal vertical stacking quality of quantum dot structures. Shown in Figure 2(a) (and representative of Q1), is a repeat of 5 layers of the InAs strain uncompensated QDs and the 10 nm spacer layer can be seen. In the z-direction, the strain induces degradation in uniformity, forming the flatter and wider QDs. This correlates to a non-ideal absorption maximum in the sub-GaAs bandgap region in addition to the dislocation-related device degradation. As seen in Figure 2(b), by the inclusion of strain compensation layers (representative of Q2, Q3 and Q4), this effect is significantly reduced. The strain compensated TEM show ordered stacks of dots with improved size control and reduced defective QD density. A representative AFM image of uncladded QDs grown on the top of the strain compensated five-layer stack is shown in Figure 2(c). Average QD size here was determined to be 7×40 nm with a dot density of $5(\pm0.5)\times10^{10}\,cm^{-2}$.

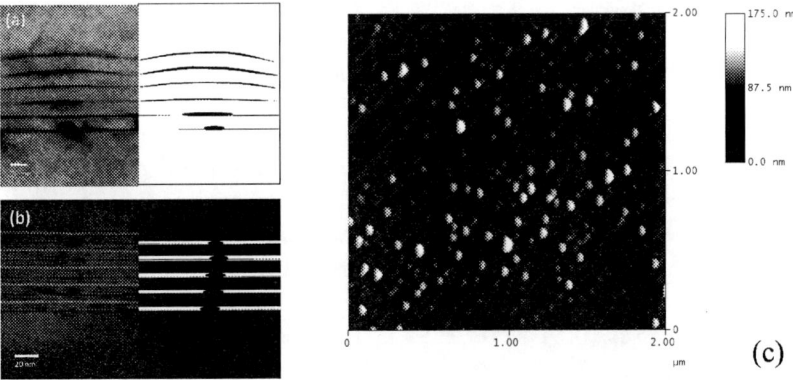

Figure 2. (a) TEM micrographs of 5x QD stacks of strain uncompensated QDs (a) exhibiting widening and flattening due to strain and well ordered stacking of QDs (b) showing no visible strain effects. (c) AFM micrograph of InAs QDs grown on top of the SC five-stack (Q2).

High resolution X-ray diffraction was also used to investigate the effects of strain compensation and superlattice periodicity. X-ray diffraction has been used previously for the characterization of strain in epitaxially grown QD superlattice structures [13, 14]. When the four samples are aligned by their GaAs peak (Figure 3), the affects of strain compensation can be seen by examining the InAs peak characteristics. The scan of Q1 with non-SC 5x QD shows weak intensity and wide FWHM at the zero order InAs peak as well as low clarity satellite peaks indicating both lower crystalline quality and degraded superlattice periodicity. The XRD results of Q2 show improved InAs peak intensity. Furthermore, scans of Q3 and Q4 show a narrower FWHM giving evidence of higher ordering in the superlattice. The Q3 and Q4, moving from an SC layer thickness of 14 Å to 18 Å, shows the InAs peak shifting towards the GaAs peak indicating a reduction in compressive strain.

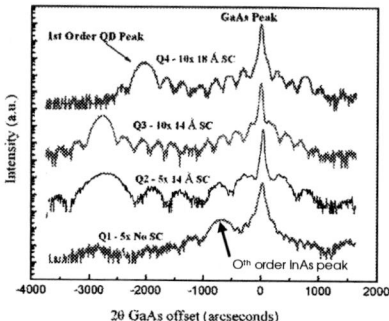

Figure 3. X-ray diffraction scans of 5x QD solar cell with no strain compensation, 5x QD test structure with 14 Å SC layer and two 10x QD test structures with SC of layer thicknesses of 14 Å and 18 Å. All of the scans have been aligned on the GaAs peak for convenience.

Solar cell results

The material quality and QD uniformity improvements seen in the materials characterization above directly affect the quantum dot solar cell device properties. The cell device results are in Table 1. The baseline solar cell, S1, is compared with the three quantum dot embedded cells in Figure 4. S3 and S4 were strain compensated with either 14Å or 18Å thick GaP layers and the final cell had no SC. Figure 4(a) shows the J-V characteristics of the four devices. The QD cell with no SC shows degradation in both I_{sc} and V_{oc} (V_{oc} = 48% of Baseline (B), and I_{sc} = 79% of B). The two cells with strain compensation, although suffering from some V_{oc} loss, exhibit marked improvement approaching 80% of baseline V_{oc}, and exhibited almost identical short circuit currents as the baseline cell. As expected, the stain accumulation in the non-SC cell leads to dislocation formation, reduced carrier lifetime and losses in current, voltage, efficiency and fill factor. Additionally, an improvement in V_{oc} can be observed moving from 14 to 18 Å of SC layer thicknesses (0.81 to 0.825 V). Fill factors are also clearly improved in the SC cells resulting from a greater shunt resistance and a lower series resistance.

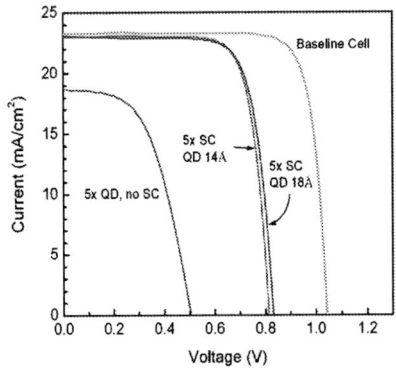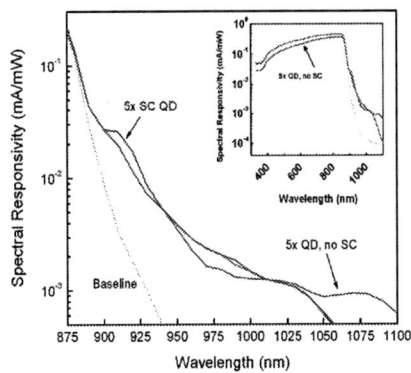

Figure 4. (a) *J-V* characteristics of a baseline, 5x QD solar cell without SC, and 5x QD solar cells with SC of 14 and 18 Å thickness. (b) Sub-GaAs bandgap spectral responsivity data of a baseline, 5x QD solar cell without SC, and two 5x QD solar cells with SC. Inset shows full spectral responsivity data of same cells.

Table 1. Solar cell results

Cell	J_{sc}	V_{oc}	FF	η
	mA/cm^2	V	%	%
S1	18.7	0.51	53.8	3.7
S2	23.1	0.81	77.3	10.7
S3	23.0	0.83	75.9	10.6
S4	23.3	1.04	85	14.5

Sub-bandgap photoconversion of a GaAs/InAs QD solar cell shows marked improvement over the baseline GaAs cell. Figure 4(b) highlights this region of the tested cells. Above the wavelength-space bandgap of GaAs, the spectral responsivity of the baseline cell drops rapidly. The cells with quantum dots show continued absorption into the near IR. A numerical integration of the product of the spectral response and the AM0 solar spectrum (with reflectance removed) can be used to calculate the short circuit current from spectral responsivity data. This was done for both wavelengths above and below the GaAs band-edge (870 nm) with the QD contributed portion of Isc being taken from 870-1100 nm. Contribution to I_{sc} in the QD samples in this region (875-1100 nm) was found to be 0.277 mA/cm^2 and 0.300 mA/cm^2 for the non-SC and SC cells respectively. The increased spectral responsivity above 1050 nm for the non-SC cell is due to the increase in QD diameter distribution in the widened QDs from layer to layer. However, the I_{sc} in the GaAs range (330-875 nm) is significantly reduced in the non-SC cell, giving an overall I_{sc} 6 mA/cm^2 less than the strain compensated cell. These results show that both non-SC and SC QD solar cells are capable of adsorption and subsequent carrier collection. However, the SC cell is able to produce both a greater QD contribution to the short circuit current in addition to recovered responsivity in the GaAs range.

CONCLUSION

The strain compensation of epitaxially grown InAs quantum dots can greatly improve solar cell device performance by improving the absorption of QDs, as well as reducing the negative effects on the solar cell I_{sc} and V_{oc} when compared with the strain uncompensated devices. TEM and AFM images confirm the effects of stress on the distortion of QDs from layer to layer; XRD analysis is consistent with this observation. Future work continues on the optimization of strain compensation of InAs quantum dots. It is anticipate that an optimal SC layer thickness, along with an increased number of dot stacks, will lead to a further increases in the QD contribution to I_{sc}. The solar cell performance improvements with SC observed in these devices are extremely encouraging for improving the current matching abilities of the GaAs middle junction and ultimately the efficiency of lattice-matched III-V triple junction solar cells.

ACKNOWLEDGEMENTS

The authors would like to thank R. Aguinaldo at the Rochester Institute of Technology for AFM images, F. Bei at RIT for XRD scans & analysis and J. Clark at RIT for solar cell fabrication. Funding for this study has been provided by NASA, AFOSR and the U.S. Department of Defense.

REFERENCES

1 D. B. Bushnell, T. N. D. Tibbits, K. W. J. Barnham, J. P. Connolly, and M. Mazzer, J. Ekins-Daukes, J. S. Roberts, G. Hill, and R. Airey, J. of Appl. Phys. 97, 124908 (2005).
2 S. Sinharoy, M. O. Patton. T. M. Valko. Sr., V.G. Weizer, Progr. In Photovoltaics: Res. & Appl. 10, 427 (2002).
3 Q. Shao, A. A. Balandin,A. I. Fedoseyev, M. Turowski, Appl. Phys. Lett. 91, 163503 (2007).
4 R.P. Raffaelle, S. Sinharoy, J. Andersen, D. Wilt, S.G. Bailey, Proc. of the IEEE World Conference on Photovoltaic Energy Conversion 1, 162-166 (2006).
5 A. Luque, A. Marti, Phys. Rev. Let. 78, pp. 5014-5017 (1997).
6 A. Marti, N. Lopez, E. Antolin, E. Canovas, C. Stanley, C. Farmer, L. Cuadra, A. Luque, Thin Solid Fims 511-512, 638 (2006).
7 C. D. Cress, S. M. Hubbard, B. J. Landi, D. M. Wilt, and R. P. Raffaelle, Appl. Phys. Lett., 91, 183108 (2007).
8 S.M. Hubbard, D. Wilt, S. Bailey, D. Byrnes, R. Raffaelle, Proc. of the IEEE World Conference on Photovoltaic Energy Conversion 1, 118 (2006).
9 S.M. Hubbard, R. Raffaelle, R. Robinson, C. Bailey, D. Wilt, D. Wolford, W. Maurer, S. Bailey, Mater. Res. Soc. Symp. Proc. 1017E, DD13-11 (2007).
10 A. Marti, N. Lopez, E. Antolin, E. Canovas, A. Luque, C. Stanley, C. Farmer, P. Diaz, Appl. Phys. Lett. 90, 233510 (2007).
11 M. Mazzer, K.W.J. Barnham, I.M. Ballard, A. Bessiere, A. Ioannides, D.C. Johnson, M.C. Lynch, T.N.D. Tibbits, J.S. Roberts, G. Hill, C. Calder, Thin Solid Films 511-512, 76 (2006).
12 S. M. Hubbard, C. D. Cress, C. G. Bailey, R. P. Raffaelle, S. G. Bailey, D. M. Wilt, Appl. Phys. Lett., submitted Nov. 2007.
13 N. Nuntawong, S. Birudavolu, C. P. Hains, S. Huang, H. Xu, and D. L. Huffaker, Appl. Phys. Lett. 85, 3050 (2004).
14 A. Krost, F. Heinrichsdorff, D. Bimberg, J. Blasing, A. Darhuber, G. Bauer, Cryst. Res. Technol. 34, 89 (1999).

Mater. Res. Soc. Symp. Proc. Vol. 1031 © 2008 Materials Research Society 1031-H08-03

TiO2 Surface State Control for Dye Sensitized Solar Cells with High Efficiency and the Solidification -Fabrication of Charge Carrier Path

Fumi Inakazu[1], Yuhei Ogomi[1], Yusuke Noma[1], Yoshihisa Fujita[1], Mitsuru Kono[2], Yoshihiro Yamaguchi[2], Yohie Kashiwa[1], Takeshi Kogo[1], and Shuzi Hayase[1]

[1]Kyushu Institute of Technology, Kitakayushu, 808-0196, Japan
[2]Nippon Steel Chemical Co. Ltd., Kitakyushu, 804-8503, Japan

ABSTRACT

Improvement of photovoltaic performance for dye sensitized solar cells (DSC) is discussed in terms of electron-path and ion-path. In order to make electron path, we focused on passivation of TiO_2 surface states which are observed by thermally stimulated current (TSC). The TiO_2 surface was well-passivated with dye molecules under pressurized CO_2 atmosphere. It was found that DSC cells prepared by a CO_2 process (Cell-CO_2) had higher efficiency than those prepared by a conventional dipping process (Cell-DIP) and the higher efficiency was associated with low TiO_2-surface state, high electron diffusion coefficient and long electron life time in TiO_2 for the Cell-CO_2. In addition, dye-staining under pressurized CO_2 atmosphere had advantages over a conventional dipping process on rapid dye-uptake and less dye aggregation. In order to fabricate ion-path in solidified electrolyte, we focused on surface modification of nano-materials. Surface of nano-materials such as TiO_2-nanoparticles and porous alumina films were modified with imidazolium iodide moieties consisting of long alkyl chains which render surface-molecules self-organized. Redox-species are concentrated on the self-organized molecules and make ion-path. We propose quasi-solid electrolyte system consisting of two layers having different charge carrier concentration to keep high photoconversion efficiencies even after solidification.

INTRODUCTION

Dye sensitized solar cells (DSCs) have attracted interest because of the high photo-conversion efficiency reaching 12 %[1-3]. The next target was set to be 15 % in order to catch up Si-based solar cells. Solidification of electrolytes or replacement of the liquid electrolyte with hole conductive materials is another important research item. In this paper, we focus on preparation of high efficiency cells and the solidification of the electrolyte. Both are associated with how to collect electrons and ions. We focus on surface modification of nano-interfaces in order to make these charge carrier passes.

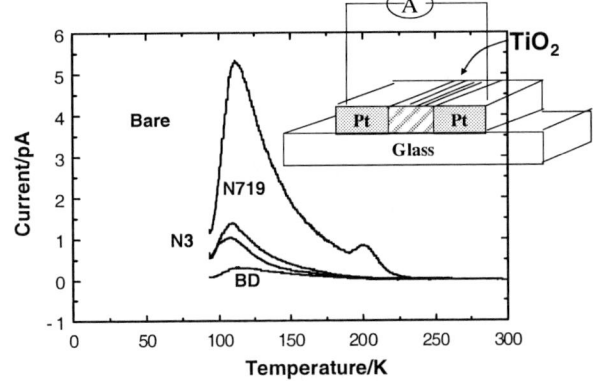

Figure 1. TSC curves for bare and stained TiO₂ layers

cis-Bis(isothiocyanato)bis(2,2'-bipyridyl-4,4'-dicarboxylato)-ruthenium(II): N3, cis-bis(isothiocyanato)bis(2,2'-bipyridyl-4,4'-dicarboxylato)-ruthenium(II) bis-tetrabutylammonium: N719, tris(isothiocyanato)-ruthenium(II)-2,2':6',2"-terpyridine-4,4',4"-tricarboxylic acid, tris-tetrabutylammonium salt: BD. TS-FETT apparatus (Rigaku model)

43

DISCUSSION

1. Fabrication of electron-path

1-1. Surface states of TiO_2 evaluated by using thermally stimulated current(TSC).

It has been reported that surface states of TiO_2 layers become electron traps and recombination centers [4]. The surface trap passivation increases electron diffusion coefficient and makes electron life time longer, resulting in increasing short-circuit current (Jsc) and open circuit voltage (Voc) respectively. It has been reported that passivation of porous TiO_2 electrodes with carboxylic acids and thin alumina layers decreases charge recombination rate and increases electron diffusion coefficients [5-15]. We evaluated these surface traps by a thermally stimulate current method (TSC) [16]. Figure 1 shows the TSC curves (relationship between current and temperature) after TiO_2 layer was stained with N3, N719 and BD. Bare nano-porous TiO_2 electrode had a large TSC peak at 125 K and a small peak at around 210 K. The trap depth of the two peaks was about 0.1 eV for the 125 K peak and 0.5 eV for the 210 K peak respectively[17]. These peaks decreased after dye-staining, suggesting that these surface traps were passivated with dye adsorption. One of the most efficient methods to inhibit charge recombination is to insert Al_2O_3 blocking layers between TiO_2 and dye. It has been reported that open circuit voltage increased by about 20-50 mV and electron life time increased by 10 times after TiO_2 layer was passivated with Al_2O_3 thin layers[7]. We also observed that 125 K-TSC peak for TiO_2/N3 substrate decreased from 17pA to 8pA after thin alumina layer was inserted between the TiO_2 layer and the N3 molecule. The results on TSC measurement was consistent with results on electron life time and electron diffusion coefficient previously reported and the TSC was found to be a useful tool to evaluate surface states of TiO_2.

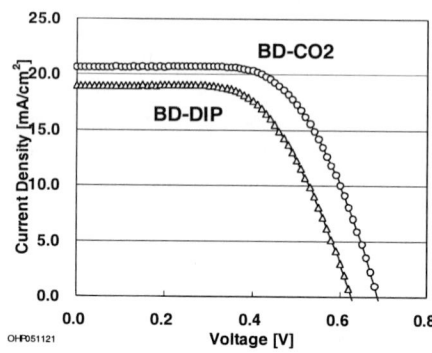

Figure 2. Photovoltaic properties for cells prepared by CO_2 and dipping processes

BD-CO2: Cell prepared by CO2 process, BD-DIP: Cell prepared by dipping process. 100mW/cm², AM1.5. LiI:500mM, TBP:580mM, I₂:50mM, MePrImI:600mM in Acetonitrile, BDCO2: 23Mpa. 40 °C, 40 min. BDDIP: 48 h at room temperature. 5 mm x 5 mm.

1-2. Dye-staining of nano-porous TiO_2 layer by pressurized CO_2 atmosphere.

In order to make TiO_2/dye substrate with high dye coverage, we used super critical CO_2 condition. It has been reported that molecular diffusion in supercritical CO_2 condition is as fast as that in gas phases[18]. In this paper, we use "pressurized CO_2" instead of "super critical CO_2" because it was difficult to determine exactly whether or not the reaction always proceeded under super critical CO_2 conditions. We have already reported apparatus and procedure for CO_2 process[19].

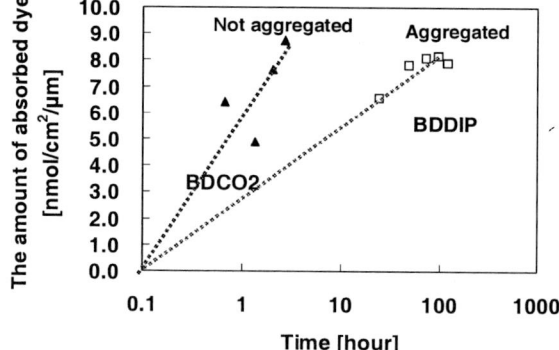

Figure 3. Relationship between the amount of dye uptake and adsorption time

BD-CO2: Cell prepared by CO2 process, BD-DIP: Cell prepared by dipping process. BDCO2: 23Mpa. 40 °C, BDDIP: at room temperature. 5 mm x 5 mm.

The DSC prepared by CO_2 process had a better Jsc and Voc than that prepared by the previous reported dipping process as shown in Figure 2. We have already reported that the electron diffusion coefficient and the electron life time in the TiO_2 layer prepared by the CO_2 process was better than that prepared by the dipping process[19].

CO_2 process shortened the dye adsorption time drastically as shown in Figure 3. It has been reported that black dye is aggregated and the aggregation decreased Voc and Jsc[20]. Black dye is commonly adsorbed with co-adsorbents such as deoxycholicacid (DCA) in order to suppress dye aggregation[20]. The aggregation was monitored by observing a 600nm absorption peak of BD/TiO_2 substrate which blue-shifted from 610nm to 585nm when the substrate was dipped for 120 h in BD solution as shown in Figure 4. We concluded that the blue shift is associated with black dye-aggregation, because the blue shift was not observed for $TiO_2/BD/DCA$. In addition, the blue-shift was not observed during the dye uptake under the

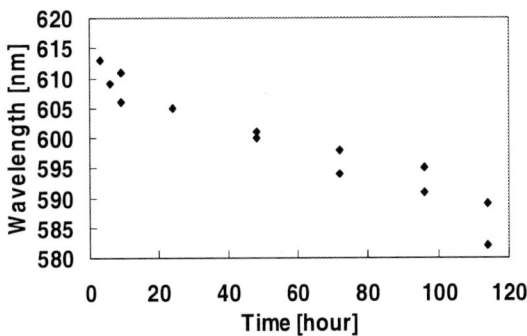

Figure 4. Relationship between light absorption maximum of BD/TiO_2 substrate and dipping time

CO_2 process, suggesting that BD is not aggregated on TiO_2/BD substrate prepared by CO_2 process.

The swift uptake of BD can be explained by swift BD diffusion into highly porous TiO_2 layer. Swift reaction of BD with TiO_2 surface is considered to be another reason. BD adsorption on a dense TiO_2 layer was monitored by UV-VIS absorption. It was found that the BD adsorption under pressurized CO_2 atmosphere was faster than that by dipping process. It has been reported that TiO_2 surface was covered with CO_2 molecules under super critical CO_2 condition and carbonate moieties are hanged from the surface[21]. The surface modified CO_2 may play an important role for the swift DB uptake. In addition, it was found that BD concentration in the pressurized CO_2 liquid was 1/10 compared with that in dipping solution. The low concentration of BD (diluted BD) in pressurized CO_2 liquid and swift reaction of BD with CO_2-modified TiO_2 surface is considered to suppress the BD aggregation. We observed 10.42% efficiency (FF 0.65, Voc, 0.70 V, Jsc 22.79 mA/cm^2, area 0.23 cm^2, AM1.5, 100mW/cm^2) for DSC consisting of TiO_2/BD substrate fabricated under pressurized CO_2

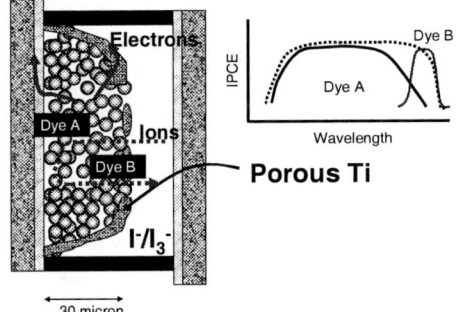

Figure 5. Three dimensional TiO_2 electrode consisting of two-dye layer and three dimensional porous Ti electrode

process without DCA. It was concluded that pressurized CO_2 is very effective for preparing electron-path in nano-porous TiO_2 layers.

1-3 Three dimensional porous Ti electrode

In order to increase photo-conversion efficiency for DSCs, sun-light with wide range of wavelength has to be adsorbed by dye molecules. DSCs stained by mixed dye solution have been already reported[22]. The cell we are aiming at is shown in Figure 5. The cell consists of a two-dye layer (A and B) and a three dimensional porous Ti or W electrode. It is very difficult to prepare the two layer structure by dipping the porous TiO_2 substrate merely in dye solution. We found that pressurized CO_2 process makes it possible. To begin with, TiO_2 substrate was stained with A dye in the pressurized CO_2 condition. At this moment, the half-top of the substrate was stained by dye A. Then, the substrate was stained by dye B which covers the rest of the unstained TiO_2 layer. We compared two type of DSC, namely, a t wo-dye layer structure (two layer type) and a mixed dye structure (cocktail type). When BD was mixed with dyes with absorption edge having more than 700 nm (cocktail type), the efficiency decreased, probably, because of the unfavorable dye aggregations or energy migrations. However, the decrease in efficiency was not observed for two layer type cells. This shows that the two layer type cell has high potential for harvesting wide range of light by combination of IR dyes with visible dyes. The details are reported elsewhere[23].

When cell thickness exceeds 30 micron, it is expected that electron collections from the top of the TiO_2 layer to the bottom TCO layers become difficult. In order to cover the insufficient electron collection, we prepared three-dimensional electrode shown in Figure 5. The problem was the thickness of Ti or W electrode because more than 150 nm thickness Ti or W films were

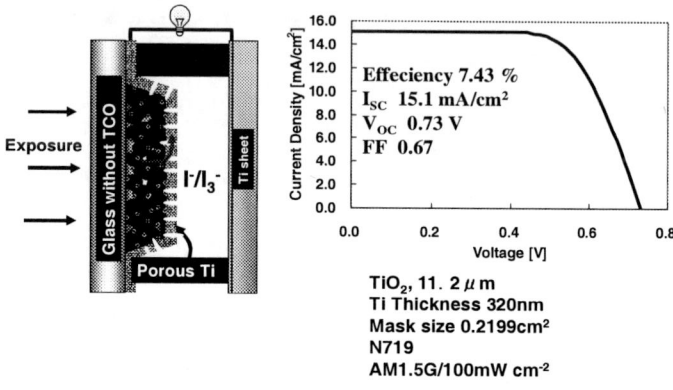

TiO₂, 11. 2 μm
Ti Thickness 320nm
Mask size 0.2199cm²
N719
AM1.5G/100mW cm⁻²

Figure 6. Photovoltaic properties for TCO-less-DSC (all metal electrode)

needed in order to obtain 10 ohm/square. Thick sputtered Ti or W electrodes covered the whole of the porous TiO_2 electrode and prevented the smooth diffusion of I_3^-/I^-. We solved the diffusion problem by using porous Ti or W three dimensional electrodes. Jsc for DSC consisting of TCO and W electrode was 17.5 mA/cm², which was higher than 16mA/cm² for DSC consisting of only TCO.

1-4 TCO-less DSC (DSC consisting of all metal electrodes)

Figure 6 shows a transparent conductive layer (TCO)-less DSC structure[24,25] consisting of our porous Ti electrode. It has been reported that TCO glass is one of the most expensive staffs. The TCO-less DSC may solve the problem. Figure 6 also shows a I-V curve and the efficiency was 7.43% which was almost the same as that for a DSC consisting of TCO glass. The results show that electrons are really collected from the

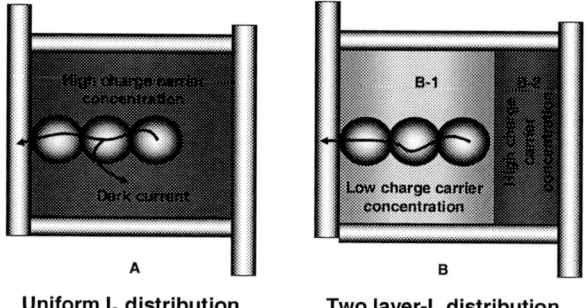

Figure 7. Concept for high efficiency quasi-solid-DSC consisting of two carrier concentration regions

Charge carrier: I⁻/I₃⁻ or electron (or hole)

three dimensional porous electrode. In addition, the porous Ti electrode opened a way to fabrication toward high-efficiency-TCO-less DSC.

2 Ion-path in quasi-solid medium

Solidification of electrolytes is one of crucial research items for DSCs[26-30]. Solidification prevents ionic diffusions and decreases photovoltaic performances for DSCs. The decrease in the photovoltaic performances becomes serious when the content of electro inactive gelators increases in order to reach all solid type DSCs. We solved the problem by fabricating ion-paths in the solidified electrolytes[31,32], where, I^-/I_3^- species are concentrated and moved by Grötthuss mechanism. Figure 7 shows the DSC structure we proposed (B). A high charge carrier concentration area is needed between TiO_2 layer and counter electrode (Figure 7 B-2). On the contrary, charge carrier concentration has to be low in nanopores of TiO_2 layer (Figure 7 B-1), otherwise, back electron transfer (dark current) increases (Figure 7 A).

In order to realize Figure 7 B structure, we used nano-particles or nano-porous Al_2O_3 film as electrochemically inactive supporter (Gelator). Appearance of the former is hard clay and the clay can be pasted on the TiO_2 layer to form the solid B-2[30]. Figure 8 shows DSC structure consisting of porous alumina films. In order to concentrate the redox species in Al_2O_3, the surface of these alumina nanopore walls was chemically modified with imidazolium iodide moiety consisting of long alkyl groups (C12 in Figure 8). The fact that I_2 is concentrated in alumina pore was observed by using adsorption experiments using Raman and UV-VIS absorption spectroscopy. Figure 8 shows photovoltaic properties for the quasi-solid DSCs consisting of the surface modified Al_2O_3 films. Jsc increased after the solidification. This is very interesting results because photovoltaic properties usually decrease after quasi-solidification.

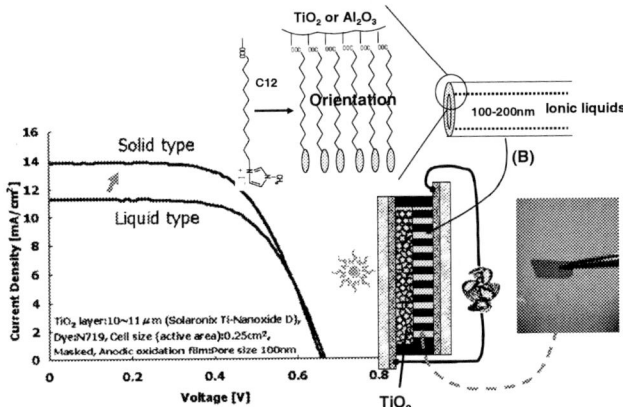

Figure 8 Photovoltaic properties for quasi-solid DSCs consisting of surface modified Al_2O_3 films

The cell consists of the surface modified porous alumina with PEDOT-PSS (conductive polymer) are shown in Figure 9. The charge carrier in the film is electron and the carrier

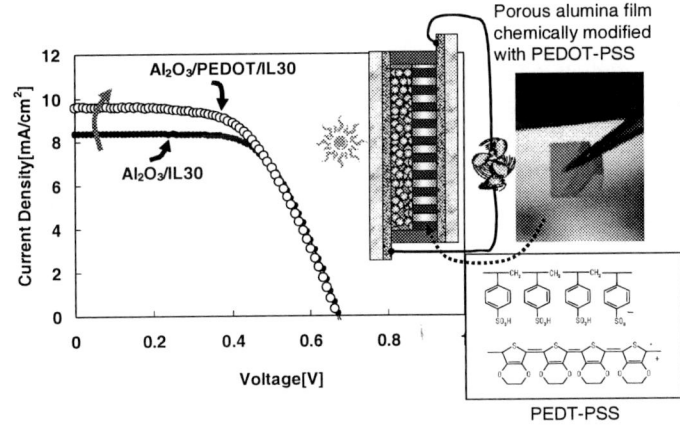

Figure 9 Photovoltaic properties for quasi-solid DSCs consisting of surface modified Al_2O_3 films with PEDOT-PSS

concentration is high in the film. Charge is carried ionically in the TiO_2 nanopores. The cell efficiency after solidification was almost the same as that before solidification. If pores in TiO_2 were filled with PEDOT-PSS, back electron transfer from the TiO_2 layer to the PEDOT-PSS occurs very swiftly and the photovoltaic property decreased. The hybrid cell consisting of ionic and electronic charge carrier layers proves again the generality of the effectiveness of the cell structure shown in Figure 8.

CONCLUSIONS

In order to improve photovoltaic properties, we focused on fabricating ion- and electron-paths. In both case, surface modifications of nano-interfaces were critical. The surface state passivation of porous TiO_2 layer with dye molecules under pressurized CO_2 atmosphere increased the electron diffusion coefficient. DSCs prepared by the CO_2 process showed better photovoltaic properties than those prepared by dipping process. The CO_2 process had advantages over a previous dipping process in terms of swift dye-uptake and suppression of dye aggregation. The cell structure consisting of two-charge-carrier-concentration was proposed and was proved to be effective for the solidification. In order to fabricate solidified media consisting of ion paths, surfaces of nano-particles or porous alumina films were modified with imidazolium moieties having a long alkyl group which facilitates molecular alignment. Hybrid charge carrier media consisting of PEDOT-PSS/porous alumina was proposed. In addition, we proposed three dimensional cell structures consisting of a two-dye layer structure and a three-dimensional porous electrode, which has potential application to high photoconversion efficiency cells.

REFERENCES

1. 1. B. O'Regan, and M. Grätzel, Nature 353, 737, 1991.
2. A. Hagfeldt and M. Graetze, Chem. Rev. 95, 347, 1995.
3. L. Han, N. Koide, Y. Chiba, A. Islam, R. Komiya, N. Fuke, A. Fukui, and R.Yamanaka, Appl. Phys. Lett. 86, 213501, 2005.
4. G. Schlichtho, S.Y. Huang, J. Sprague, A.J. Frank, J. Phys. Chem. B 101, 8141, 1997.
5. S. Nakade, Y. Saito, W. Kubo, T. Kanzaki, T. Kitamura, Y. Wada and S. Yanagida, Electrochem. Commun. 5, 804, 2003.
6. S. Sakaguchi, H. Ueki, T. Kato, T. Kado, R. Shiratuchi, W. Takashima, K. Kaneto, and S. Hayase, J. Photochem. Photobiol. A. Chem. 164, 117, 2004.
7. F. Fabregat-Santiago and J. García-Cañadas, E. Palomares, J. N. Clifford, S. A. Haque, and J. R. Durrant, J. Appl. Phys. 96, 6903, 2004.
8. G. R. R. A. Kumara, K. Tennakone, V. P. S. Perera, A. Konno, S. Kaneko, and M. Okuya, J. Phys. D. 41, 868, 2001.
9. A. Zaban, S. G. Chen, S. Chappel, and B. A. Gregg, Chem. Commun. 22, 2231, 2002.
10. S. Chappel, S. G. Chen, and A. Zaban, Langmuir 18, 3336, 2002.
11. Y. Diamant, S. G. Chen, O. Melamed, and A. Zaban, J. Phys. Chem. B. 107, 1977, 2003.
12. E. Palomares, J. N. Clifford, S. A. Haque, T. Lutz, and J. R. Durrant, Chem. Commun. 14, 1464, 2002.
13. E. Palomares, J. N. Clifford, S. A. Haque, T. Lutz, and J. R. Durrant, J. Am. Chem. Soc. 125, 475, 2003.
14. F. Lenzmann, M. Nanu, O. Kijatkina, and A. Belaidi, Thin Solid Films 451, 639, 2004.
15. Y. Fukai, Y. Kondo, S. Mori, E. Suzuki, Electrochem. Commun. 9, 1439, (2007).
16. T. P. Nguyen, Materials Science in Semiconductor Processing 9, 198, 2006.
17. A. Opanowics and P. Petrucha, J. Appl. Phys. 93, 957, 2003.
18. P. G. Jessop, T. Ikariya, R. Noyori, Nature, 368, 231, 1994.
19. Y. Ogomi, S. Sakaguchi, T. Kado, M. Kono, Y. Yamaguchi, S. Hayase, J. Electrochem. Soc., 153, A2294, 2006.
20. L. M. Peter, and K. G. U. Wijayatha, Electrochim. Acta., 45, 4543, 2000.
21. W. Gu and C. P. Tripp, Langmuir 22, 5748, 2006
22. H. Otaka, M. Kira, K. Yano, S. Ito, H. Mitekura, T. Kawatam F. Matsui, J. Photochem. Photobio. A: Chem., 164, 67, 2004.
23. Y. Ogomi, S. Sakaguchi, S. Hayase, Proceeding of SPIE, 2007 (San Diego)
24. J. M. Kroon1 et al., Prog. Photovolt: Res. Appl., 13 333, 2005.
25. N. Fuke, A. Fukui, Y. Chiba, R. Komiya, R. Yamanaka, and L. Han, Jpn. J. Appl. Phys., 46, L420, 2007.
26. K. Tennakone, G. K. R .Senadeera, V. P. S. Perera, I. R. M. Kottegoda, L. A. A. De silra, Chem. Mater.11, 2474, 1999.
27. P. Wang, S. M. Zakeeruddin, J. Moser, M. K. Nazeeruddin, T. Sekiguchi, M. Gratzel, Nat. Mater. 2, 402, 2003.
28. W. Kubo, K. Murakoshi, T. Kitamura, S. Yoshida, M. Haruki, K. Hanabusa, H. Shirai, Y. Wada, S. Yanagida, J. Phys. Chem.B 105, 12809, 2001.
29. P. Wang, S. M. Zakkeruddin, I. Exnar, M. Gratzel, Chem. Commun. 2972, 2002.
30. P. Wang, S. M. Zakkeruddin, P. Comte, I. Exnar, M. Graetzel, J. Am. Chem. Soc. 125, 1166, 2003.
31. T. Kato, T. Kado, S. Tanaka, A. Okazaki and S. Hayase, J. Electrochem. Soc. 153, A626, 2006.
32. T. Kato and S. Hayase, J. Electrochem. Soc. 154, B112, 2007.

Mater. Res. Soc. Symp. Proc. Vol. 1031 © 2008 Materials Research Society 1031-H09-21

Polaron Pair Dissociation and Polaron Recombination in Polymer: Fullerene Solar Cells

Carsten Deibel, Andreas Baumann, Jens Lorrmann, and Vladimir Dyakonov
Experimental Physics VI, Julius-Maximilians-University of Wurzburg, Am Hubland, Wurzburg, 97074, Germany

ABSTRACT

In polymer:fullerene solar cells, the field-dependent photocurrent is due to a combination of polaron pair dissociation, competing with monomolecular recombination, and bimolecular recombination. We compare of the experimental photocurrent of P3HT:PCBM bulk heterojunction solar cells to Monte Carlo simulations of bilayer and bulk heterojunction devices. The shape of the photocurrent can be reproduced, the high quantum yield, however, cannot be explained with help of the simulations. In order to analyse the dominant experimental recombination mechanism, we apply the photo-CELIV technique. Our data can be fitted with bimolecular recombination, but only if a reduced Langevin recombination factor is assumed. Thermalisation is accounted for by measuring the time-dependent charge carrier mobility.

INTRODUCTION

Organic bulk heterojunction solar cells have received a growing interest of the scientific community during the last years. A lot of research has been dedicated to the optimisation as well as the gaining of an enhanced physical understanding of these devices; significant progress has been made in both directions [1, 2]. Despite these efforts, the elementary processes of carrier generation, transport and extraction are not fully understood yet. Unresolved issues concern the high polaron pair dissociation yield and the dominant recombination mechanism of polarons in bulk heterojunction solar cells. For bilayer solar cells, the strong influence of the monomolecular process of polaron pair dissociation on the photocurrent as well as the solar cell fill factor has been shown by Peumans et al. [3], highlighting its relevance for organic photovoltaics. In bulk heterojunction polymer solar cells, the disorder induced by the two material components and their mixture leads to favorable exciton dissociation, but more difficult polaron pair dissociation and polaron transport. Usually, bimolecular processes are seen experimentally by charge extraction and photoinduced absorption [4, 5], highlighting the polaron pair dissociation yield. In this contribution, we address the mechanism and efficiency of geminate and nongeminate recombination of polymer–fullerene solar cells by comparing experiments and Monte Carlo simulations of photocurrent and time-resolved charge extraction.

EXPERIMENTAL DETAILS

We prepared organic bulk heterojunction solar cells by spin coating 1:1 blends of poly[3-hexyl thiophene-2,5-diyl] (P3HT) with [6,6]-phenyl-C_{61} butyric acid methyl ester (PCBM), 2wt% dissolved in Chlorobenzene, on PEDOT:PSS covered ITO/glass substrates. The active layer was around 300nm thick, and was annealed after spin coating. Al anodes were thermally evaporated. We obtained P3HT from Rieke Metals and PCBM from Solenne.

Temperature-dependent current–voltage measurements were performed, in darkness and under illumination (typically at 100mW/cm^2). In order to obtain the photocurrent vs. the internal voltage, the voltage at the crossing point of dark and illuminated current–voltage curve, V_0, was taken as approximation of zero internal field. We note that this method slightly underestimates the contact potential difference, the so called built-in potential.

A versatile experimental technique for investigating charge transport in these disordered materials is the photo-CELIV method, charge extraction by linearly increasing voltage [6]. Charge carriers are generated by a short laser pulse (Nitrogen laser with dye unit, ~5ns, 30μJ/cm^2). After a certain delay time at zero internal field, we dissociate a fraction of the polaron pairs and extract these charges by a triangular voltage. Thus, the charge carrier mobility from the maximum of the extraction peak, and the concentration of extracted carriers can be obtained simultaneously. The dependence on the delay time gives us information on the carrier thermalisation (mobility) as well as the recombination mechanism (concentration).

Monte Carlo simulations of hopping transport in a gaussian density of states within a cubic lattice were done for comparison. Coulomb interaction of the charge carriers as well as geminate (and, partly, nongeminate) recombination were accounted for. The results were averaged over 5000 simulation runs, with one polaron pair per run if not stated otherwise; i.e., mostly only geminate (monomolecular) recombination was considered. We simulated the photocurrent (neglecting current injection through the electrodes) as well as photo-CELIV experiments.

DISCUSSION

Photocurrent & Polaron Pair Dissociation Yield

In Fig. 1(a), the polaron pair dissociation yield in dependence on the internal electric field was simulated for bilayer and bulk heterojunction solar cells. Also shown is the experimental, normalised photocurrent of a bulk heterojunction — please note that it might include bimolecular recombination in addition to pure monomolecular decay. The 50–90% quantum yield observed in good organic solar cells can only be achieved in the simulations, if an internal field of about 10–30 times the field under short-circuit conditions of a working bulk heterojunction solar cell is applied. This effect could be explained by extremely long polaron pair lifetimes, which is unlikely. For spatially ordered heterojunctions such as bilayers, the efficient polaron pair dissociation can be in part explained by assuming that the major part of the voltage drops over a narrow depletion region at the heterojunction [3]. In contrast, for a spatially distributed bulk heterojunction, the low dielectric constant does not allow for auch a locally enhanced electric field, rendering the question of high yield at moderate fields difficult to solve. Some possible explanations include the dark dipole model by Arkhipov et al. [7], which, in analogy to the locally enhanced electric field, should work best if a certain degree of order is attained for avoiding cancelling of the dipole effect. Another explanation is hot exciton dissociation yielding a higher polaron pair interpair radius, thus enhancing dissociation. This explanation was rejected by Arkhipov et al. We propose a scenario based on the notion that a large electron–hole mobility ratio [3] or a large ratio of the donor to acceptor disorder parameter [8] enhances the polaron pair dissociation yield. We hold the fast intrachain transport of the positive polaron within the

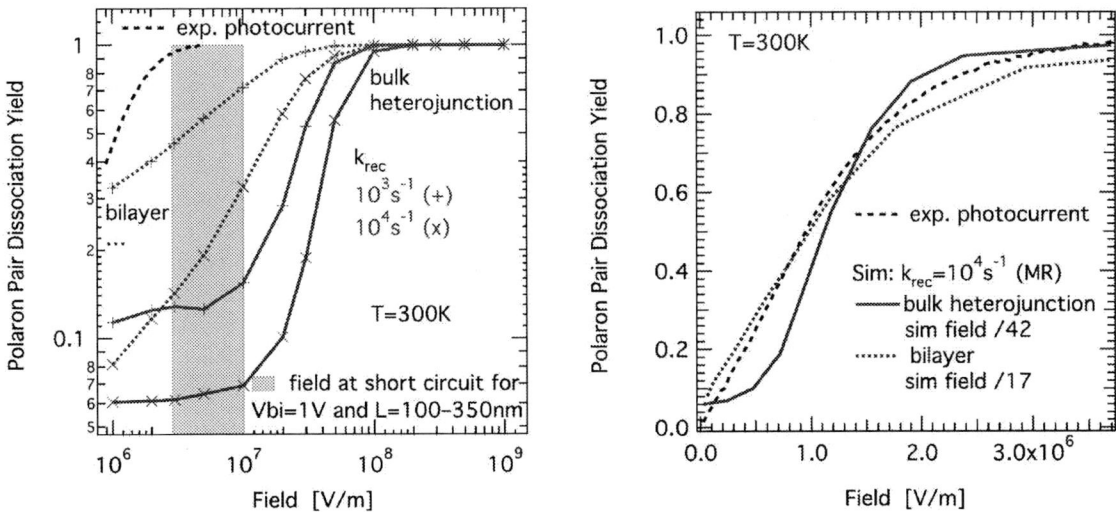

Figure 1. (a) Monte Carlo simulation results of polaron pair dissociation yield in dependence on internal electric field for bilayer (dotted line) and bulk heterojunction devices (solid line). The photocurrent of a P3HT:PCBM bulk heterojunction solar cell — including bimolecular recombination — is shown for comparison (dashed line). Note the discrepancy between the experimentally measured yield and the simulation results. (b) Polaron pair dissociation yield vs internal field on a linear scale. The electric fields of the simulations were reduced by constant factors in order to compare the shapes of polaron pair dissociation yield to the experimental photocurrent.

conjugated segment of a polymer chain, in comparison to the slower electron hopping transport from fullerene to fullerene, for the relevant mechanism enhancing the quantum yield.

Disregarding the unexpectedly high yield, we consider the field-dependence of the photocurent. As shown in Fig. 2(b), rescaling the fields used in the simulations with constant factors allows the comparison of the shape of the simulated and experimental yield curves — particularly in the high yield regime, which is less prone to errors in the calculation of the experimental internal field, as well as contributions from the field-independent polaron pair dissociation in the vicinity of the contacts. The photocurrent lies between the simulation results of bilayer and bulk heterojunction. This can be attributed to the annealed P3HT:PCBM solar cell inhibiting a certain degree of donor–acceptor phase separation.

Recombination Mechanism

We now focus on gaining insight into the dominant recombination mechanism by applying time-resolved photo-CELIV on the samples. An exemplary measurement is shown in Fig. 2(a). The carrier concentration vs delay time, and the mobility during thermalisation (see inset) is shown in Fig. 2(b) for 150K temperature. Thermalisation is observed, in contrast to

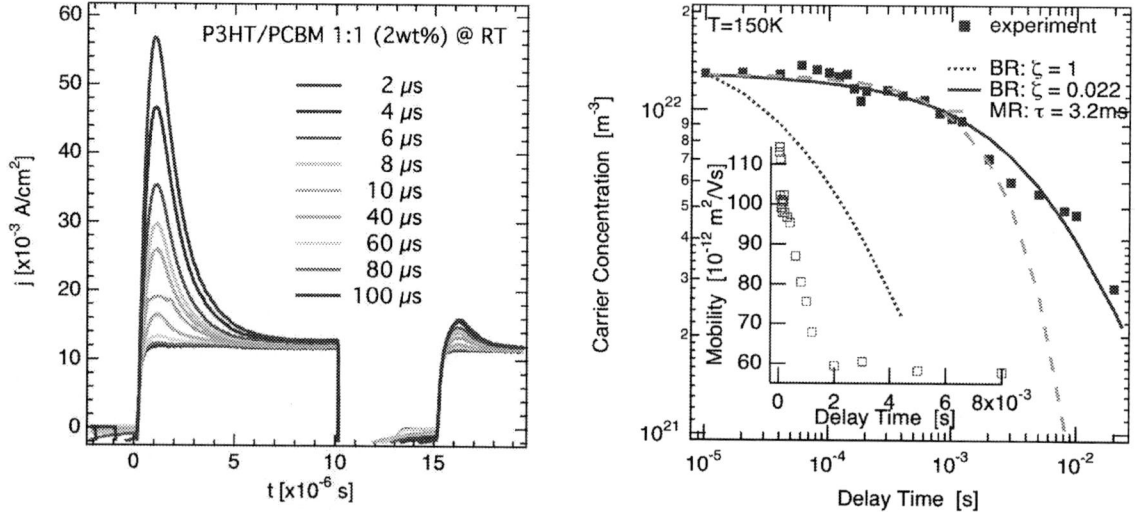

Figure 2. (a) Photo-CELIV measurement of a P3HT:PCBM bulk heterojunction solar cell with variable delay time at 300K. The second set of extraction peaks after 15µs signifies a very low charge carrier injection at the electrodes (b) Extracted carrier concentration and mobility (see inset) vs delay time. See text for an explanation.

room temperature measurements (for delay times up to 10ms). We point out that the experimental photo-CELIV signals at 150K and room temperature can neither be fitted by pure monomolecular, nor by pure bimolecular recombination. For the latter, the mobility — including thermalisation, which leads to a less steep decline of the carrier concentration with delay time — was taken into account. We used the following rate equation for describing the CELIV data:

$$\frac{dn}{dt} = G_0 - \frac{n}{\tau} - \gamma np$$

Here, n is the electron and p the hole concentration in dependence of time t, G_0 is the generation rate for charge carriers. γ is the Langevin recombination parameter, which is linearly proportional to the charge carrier mobility. Thus, for monomolecular recombination, the polaron pair lifetime τ is the fitting parameter. In contrast, bimolecular recombination does not have a free parameter, as the time delay-dependent mobility is determined experimentally. However, the data can only be fitted using bimolecular recombination by assuming a time-independent, temperature-dependent fitting prefactor ζ, which reduces the Langevin factor, and thus the mobility, to a tenth or a hundredth of the experimentally determined value. Therefore, only bimolecular recombination with a reduced mobility can describe our CELIV experiments. We note that Mozer et al. [5] also apply a fitting factor by using stretched exponentials for including the effect of thermalisation, but without comparison to their experimentally determined mobility. However, as the mobility data includes the information on thermalisation, and the impact of a linear scaling factor is easier to consider, we prefer the approach presented here.

In Fig. 3, experimental photo-CELIV curves are compared to Monte-Carlo simulated ones, again only considering monomolecular recombination. Experimentally, the carrier concentration at 300K is much more prone to recombination as compared to 150K. The

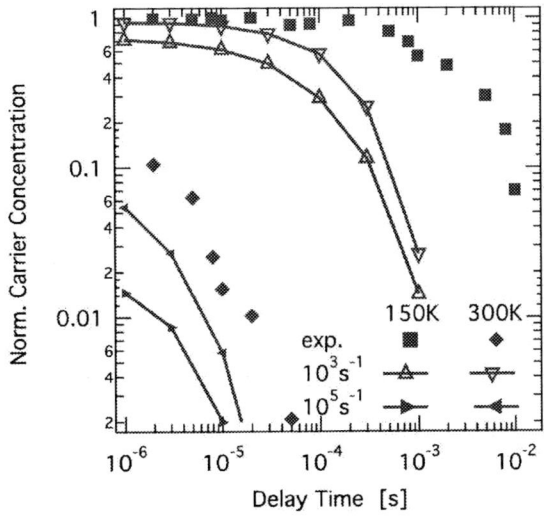

Figure 3. Experimental photo-CELIV curves measured on a P3HT:PCBM sample are compared to Monte Carlo simulations, the latter again only considering monomolecular recombination. Despite the similar shape of experimental and simulated curves, the temperature dependence of the experimental data indicates strongly indicates bimolecular recombination.

monomolecular recombination applied in the Monte Carlo simulations — although roughly fitting the shape — clearly cannot reproduce the temperature dependence of the experimental curves. The latter are limited by bimolecular recombination, particularly at room temperature. By performing Monte Carlo simulations with trapped charges at a concentration corresponding to $2 \times 10^{22} m^{-3}$, we observe a rising fraction of bimolecular recombination as well, increasing with temperature (not shown). However, in order to reach a similar importance of bimolecular recombination as compared to the experiments, we estimate that at least a 10x higher carrier concentration would be needed, clearly exceeding the experimentally determined carrier concentration.

CONCLUSIONS

We presented experiments and simulations highlighting two discrepancies related to polaron pair dissociation and polaron recombination in state-of-the-art polymer:fullerene solar cells. The experimental quantum yield of P3HT:PCBM bulk heterojunction solar cells is unexpectedly high and cannot be reproduced by Monte Carlo simulations featuring dissociation of Coulomb bound polaron pairs. We postulate that fast intrachain hopping on the polymer chain allows for the high polaron pair dissociation yield. We also investigate the dominant recombination mechanism by measuring time-dependent carrier concentration and mobility simultaneously, using the photo-CELIV technique. The characteristic temperature dependence as well as the shape evidence bimolecular recombination behaviour. Accounting for carrier thermalisation, we had to assume a reduced Langevin recombination factor in order to quantitatively fit our experimental data.

ACKNOWLEDGMENTS

C.D. acknowledges financial support of the Deutsche Forschungsgemeinschaft. A.B. acknowledges funding by the Deutsche Bundesstiftung Umwelt.

REFERENCES

1. C. J. Brabec, V. Dyakonov, J. Parisi, and N. S. Sariciftci, eds., Organic Photovoltaics (Springer, Berlin, 2003)
2. H. Hoppe and N. S. Sariciftci, J. Mater. Res. 19, 1924 (2004)
3. P. Peumans and S. R. Forrest, Chem. Phys. Lett. 398, 27 (2004)
4. M. Wohlgenannt, Phys. Stat. Sol. A 201, 1188 (2004)
5. A. J. Mozer, G. Dennler, N. S. Sariciftci, M. Westerling, A. Pivrikas, R. Österbacka, and G. Juska, Phys. Rev. B 72, 035217 (2005)
6. G. Juska, K. Arlauskas, and M. Viliunas, and J. Kocka, Phys. Rev. Lett. 84, 4946 (2000)
7. V. I. Arkhipov, P. Heremans, and H. Bässler, Appl. Phys. Lett. 82, 4605 (2003)
8. T. Offermans, S. C. J. Meskers, R. A. J. Janssen, Chem. Phys. 308, 125 (2005)

Mater. Res. Soc. Symp. Proc. Vol. 1031 © 2008 Materials Research Society   1031-H09-23

Environmental Passivation and Temperature Cycling of PCBM - Polymer Solar Cells

Annick Anctil[1,2], Andrew Merrill[2], Cory Cress[1,2], Brian Landi[2], and Ryne Raffaelle[1,2]

[1]Microsystems Engineering, Rochester Institute of Technology, Rochester, NY, 14623

[2]Nanopower Research Laboratories, Rochester Institute of Technology, 85 Lomb Memorial Drive, Rochester, NY, 14623

ABSTRACT

In the present work, polymer solar cells were fabricated from composite blends of poly[2-methoxy-5-(2-ethylhexyloxy)-1,4-phenylenevinylene] (MEH-PPV) and poly(3-hexylthiophene)-(P3HT with PCBM[60] and PCBM[70]. The composite blends were used as active layers in an ITO/PEDOT:PSS/active layer/Al device structure. Power conversion efficiencies have been measured from current-voltage (*I-V*) measurements for each of these different composite blends under simulated AM1.5 illumination. In the case of the MEH-PPV devices, the *I-V* performance has been measured as a function of polymer molecular weight, type of fullerene derivative (C_{60} or C_{70}), and PCBM:polymer ratios. The highest efficiencies for the ranges used in this study were obtained using the 150,000 g/mol MEH-PPV molecular weight, the C_{70} PCBM derivative, and a 1:4 MEH-PPV:PCBM ratio. The effect of thermal cycling on the *I-V* performance for both MEH-PPV and P3HT devices has also been measured from 77K to 330K. The devices exhibited a positive temperature coefficient for the short-circuit current density (J_{sc}), which dominated the overall efficiency of the device over this temperature range. Finally, the use of a combination of parylene and polymethylmethacralate for device passivation was shown to provide a dramatic reduction in device degradation under ambient conditions as compared to non-passivated devices.

INTRODUCTION

Recent progress with P3HT devices has resulted in efficiencies of approximately 5%[1], although concerns over limitations due to material properties (i.e. bandgap, mobility, etc.) promote alternative strategies. The use of MEH-PPV as an active layer has shown promise, although with limited efficiencies to date [2].

In devices incorporating P3HT, it has been shown that higher molecular weight (MW) P3HT resulted in slower degradation and increased glass transition temperature (Tg) [3]. This result will obviously affect the usable temperature range of these devices. Faster degradation is expected to occur at a temperature close to the Tg due to the increased motion of the polymer chains and possible diffusion of electrode degradation products [4]. Comparable studies of MEH-PPV blends are presently lacking, but may offer advantages in terms of environmental stability and power conversion efficiency.

Conventional polymer devices generally employ a PEDOT:PSS layer and a derivatized polythiophene or polyphenylenevinylene (PPV) composite blend sandwiched between two electrodes. It is well established that exposure to ambient atmosphere will degrade device performance on the order of hours to a few days. Practical implementation of these devices will rely upon the development of efficient passivation designs to maintain performance under typical environmental conditions: thermal gradients (-20 °C to 50 °C), humidity, and oxygen.

Various techniques have been developed in the last few years for polymer device encapsulation, but none of them offer a simple, permanent solution. For example, encapsulation using glass plates and epoxy is complex and requires curing of the epoxy at high temperature (80 °C to 100 °C) which may degrade the polymer [5]. Recently, alternative encapsulation schemes have emerged, incorporating alternating layers of organic and inorganic materials (e.g. aluminum oxide or silicon oxide).[6] The organic material prevents moisture degradation while the inorganic material acts as an efficient barrier against oxygen and significantly reduces the effect of pin-holes in the organic material.

In an attempt to improve MEH-PPV:PCBM device efficiencies, the effects of molecular weight and fullerene derivative (type and quantity) were investigated. The effects of thermal cycling on the *I-V* measurements for the MEH-PPV:PCBM composites were compared to those of P3HT:PCBM devices between 77K and 330K. To overcome complex encapsulation techniques, a passivation approach was evaluated using two other polymer materials: parylene and polymethylmethacralate (PMMA). Parylene (used as the initial layer) provides a pin-hole free coating at room temperature which is excellent for corrosion resistance and moisture protection. However, due to the oxygen permeability in parylene, a second layer using PMMA was employed to remedy this deficiency and prevent oxygen degradation.

EXPERIMENT

Bulk heterojunction (BHJ) solar cells were fabricated following the sandwich structure ITO/PEDOT:PSS/active layer/Al. Glass coated ITO substrates were cleaned by ultrasonication in acetone and isopropyl alcohol and dried at 120 °C for 10 minutes. A layer of poly(3,4-ethylenedioxythiophene):poly(4-styrene sulfonate) (PEDOT:PSS) was spin-coated at 4000 rpm to obtain a 60 nm thick hole conductive layer which was annealed at 120 °C for 1 hour. Solutions of polymer and C_{60} PCBM or C_{70} PCBM were mixed in *o*-xylene until a proper dispersion was obtained. Three MEH-PPV molecular weight purchased from Sigma-Aldrich were investigated: low (Mn=40,000-70,000), medium (Mn=70,000-100,000) and high (Mn=150,000-250,000). The P3HT (Rieke Metals) composites were prepared at 25 mg/mL concentrations using 1,2-dichlorobenze solutions and a 1:1 PCBM:polymer ratio. The composite solutions were spin-coated such as to obtain a 100 nm thick blend layer. The back electrical contact was 1000 Å of evaporated aluminum with an area of 0.08 cm^2.

For the passivation study, electrical contacts were prepared by wire bonding to allow I-V measurements after encapsulation. Deposition of the parylene C dimer was performed using a parylene coating system. The PMMA solution in tetrahydrofuran (*THF*) was deposited on the device coated with parylene. The whole device was dried at 80 °C for 10 minutes under argon. The *I-V* curves for all devices were obtained using an Agilent source-measure unit and a Newport Oriel Instrument light source calibrated under air mass 1.5 (AM1.5). The thermal cycles was performed using an Oxford liquid nitrogen cryostat with temperature control. Temperature dependent *I-V* data were collected from 77 K to 330 K. In order to minimize variation between data sets, comparisons were made only between devices prepared in parallel.

DISCUSSION

As shown in Figure 1, for a constant MEH-PPV:PCBM ratio (1:4), the open circuit voltage (*V$_{oc}$*) and the fill factor (*FF*) consistently increased with increasing molecular weight.

This effect results in a doubling of the overall efficiency compared to the lowest MW combination (see Figure 1b).

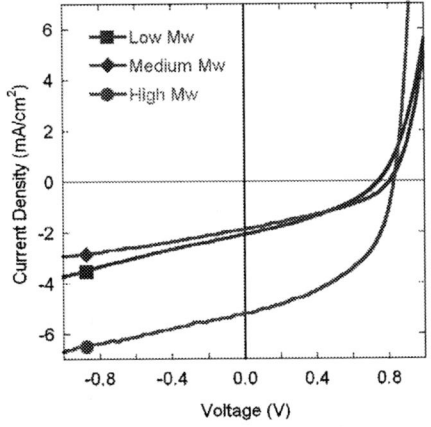

Mw	Jsc (mA/cm²)	Voc (mV)	FF (%)	η (%)
Low Mw: 40,000 - 70,000	2.07	747	36.0	0.57
Medium Mw: 70,000 - 100,000	1.88	808	37.4	0.57
High Mw: 150,000 - 250,000	3.89	835	38.5	1.25

(a)   (b)

Figure 1:(a) *J-V* characterization of BHJ MEH-PPV:PCBM (1:4) blends made with different molecular weights and (b) calculated *FF* and efficiency

This improved efficiency with increased molecular weight is consistent with what was reported for P3HT [3]. It has been suggested to be a result of longer conjugation length and better electrical interconnections within the BHJ [7]. The effect of high molecular weight on performance may have practical limitations due to the decreasing solubility of the polymer at high MW.

The effect of fullerene on the aforementioned devices was studied using the high MW polymer and a 1:4 polymer:PCBM ratio. Using PCBM C_{70} instead of C_{60} had a significant impact on the J_{sc} but provided no noticeable change on the V_{oc}. The J_{sc} increased by 33% which resulted in an absolute efficiency improvement of 0.81%.

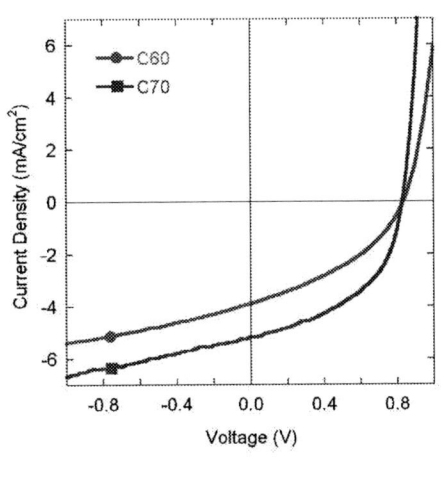

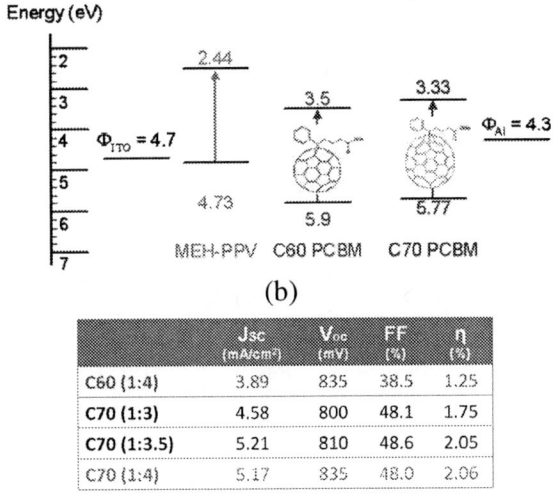

(b)

	Jsc (mA/cm²)	Voc (mV)	FF (%)	η (%)
C60 (1:4)	3.89	835	38.5	1.25
C70 (1:3)	4.58	800	48.1	1.75
C70 (1:3.5)	5.21	810	48.6	2.05
C70 (1:4)	5.17	835	48.0	2.06

(a)   (c)

Figure 2: Effect of fullerene derivative on *J-V* measurements in MEH-PPV: PCBM (1:4) for high Mw MEH-PPV and energy level diagram for the bulk heterojunction components.

57

According to the previously reported model [8], the V_{oc} should be related to the difference between the HOMO level of the donor and the LUMO of the acceptor. However, based on the cyclic voltammetry measurements performed in our laboratory [9] (Figure 2b), the lower LUMO of the C_{70} as compared to C_{60}, should have resulted in an increases in the V_{oc}. The lack of change in V_{oc} with fullerene is in contrast to P3HT blends, and undoubtedly relates to the differences in interaction between these polymers. The 1:4 weight ratio for MEH-PPV–PCBM[60] devices is considered standard for this type of device [2], but has not been verified for other derivatives such as C_{70}.

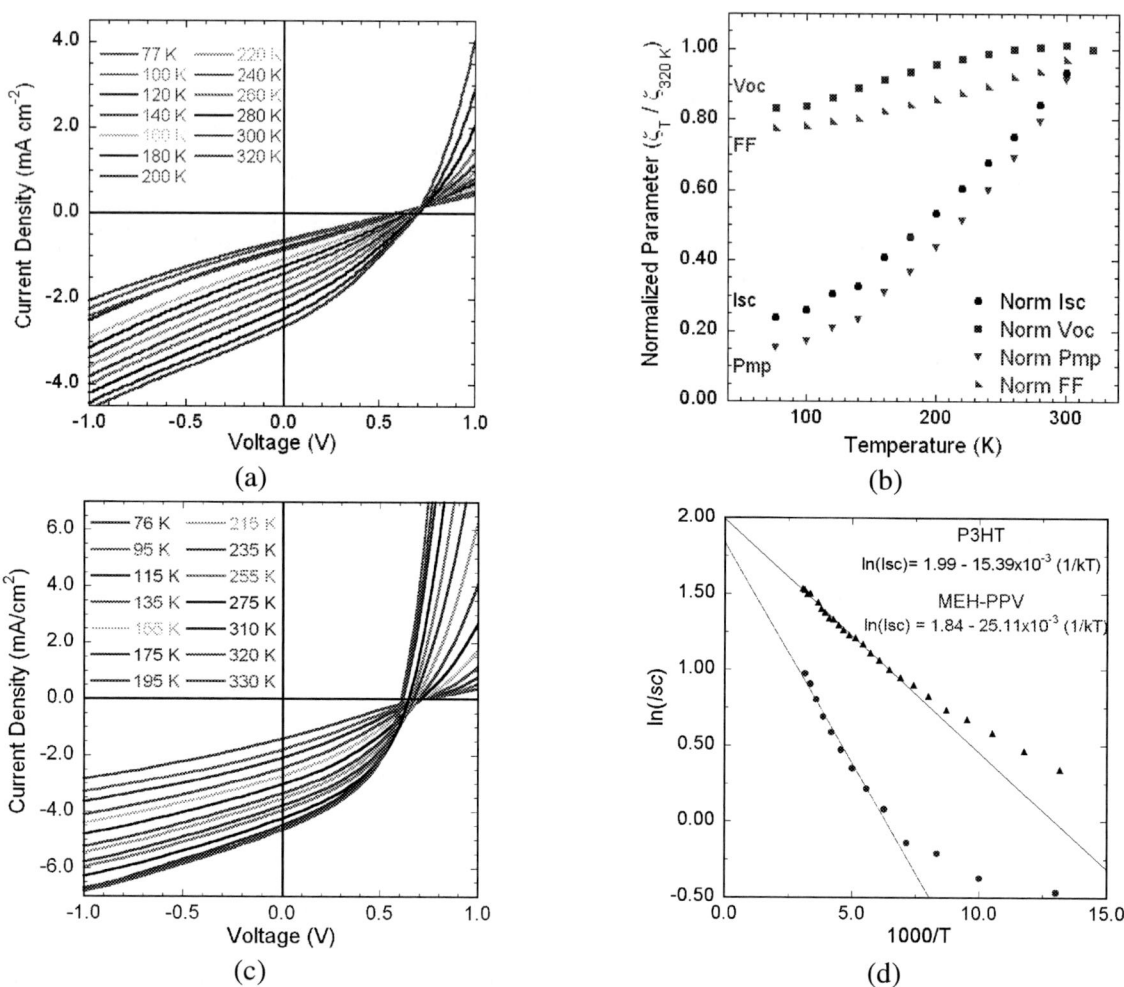

Figure 3: (a) *J-V* characterization as a function of temperature under vacuum between 77K - 320 for high Mw MEH-PPV:PCBM (1:4) (b) Effect of temperature *on Voc, FF, J_{sc} and P_{mp}* for MEH-PPV:PCBM (c) *I-V* for P3HT:PCBM (1:1) and (d) ln(*Isc*) for P3HT and MEH-PPV

In a side-by-side comparison, the effects of the MEH-PPV:PCBM[70] weight ratio on device performance has been measured for PCBM amounts modulated between 1:3 to 1:4 (see Figure 2c for data). The J_{sc} was significantly lower for a 1:3 ratio, but remained stable for both the 1:3.5 and 1:4 ratios. The V_{oc} steadily decreased with decreasing MEH-PCBM ratio. The increasing PCBM ratio led to a red-shift in the spectral response measured band edge, which is significant considering that organic solar cells performances are mainly limited by their low photo-conversion in the near-infrared.

Since charge transport in organic solar cells is affected by temperature, thermal cycling was performed in the range of 77K to 330 K for MEH-PPV and compared to the performance of un-annealed P3HT:PCBM devices. For MEH-PPV (Figure 3a), the current density and the open circuit voltage increase almost linearly with temperature as shown in Figure 3b. The increased efficiency with temperature is mostly attributed to the increased current density. For P3HT (Figure 3c), the current density follows the same trend has the MEH-PPV devices but the open circuit voltage does not. This V_{oc} decrease may be attributed to an annealing effect since the P3HT was not pre-annealed. Figure 3d contains an Arrhenius plot of the J_{sc} for the MEH-PPV and P3HT devices. A variation in activation energy (i.e., 25.1 meV vs. 15.4 meV) was observed for the MEH-PPV and P3HT devices, which is consistent with previously reported values [10],[11].

The affects of a novel passivation scheme on the J_{sc} and FF were investigated using P3HT:PCBM[60] devices. The initial reduction in J_{sc} in the parylene and parylene+PMMA devices, shown in Figure 4, is attributed to longer ambient exposure during transport to the parylene coater. The parylene+PMMA device was annealed at high temperature to dry the PMMA which may have caused its beginning of life J_{sc} to be reduced even further. The J_{sc} of the non-passivated device degrades rapidly at a rate of 0.25% normalized to the beginning of life value, with full degradation occurring after 70 hours. The parylene coated device shows a much slower rate of degradation in J_{sc} and FF: 80% of their beginning of life parameter values were maintained after 70 hours. The device coated with the novel parylene+PMMA passivation scheme showed improved J_{sc} and FF after 70 hours.

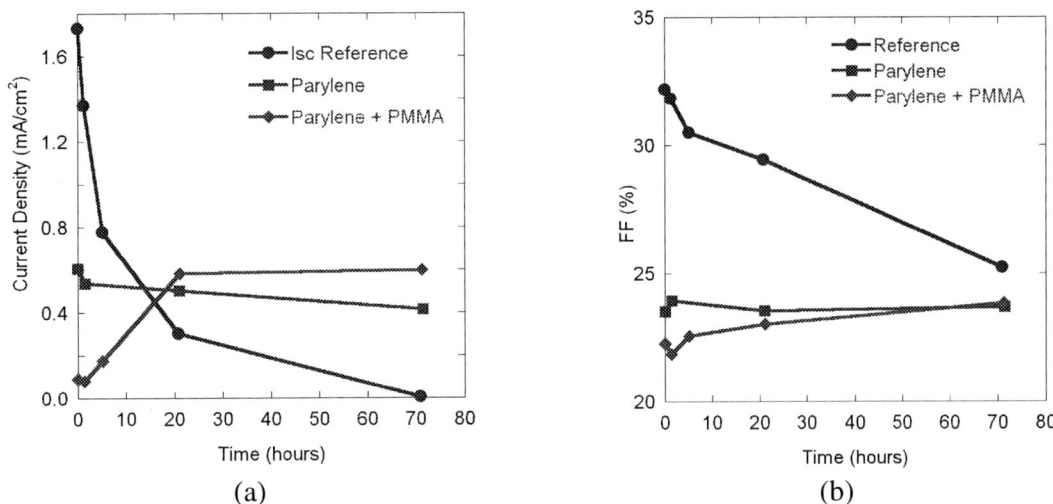

Figure 4: (a) *Jsc* as a function of time for a non passivated device, a device protected by parylene only and a device protected by parylene and PMMA (b) Fill factor as a function of time for the same 3 devices

CONCLUSIONS

Efficiencies of 1:4 MEH-PPV:PCBM were improved from 0.57% to 2.06% under AM1.5 by increasing the polymer molecular weight and the replacement of C_{60} with C_{70} PCBM in a BHJ device structure. Increasing the molecular weight of MEH-PPV:PCBM had a beneficial impact on the overall device performance. Using C_{70} PCBM instead of the C_{60} PCBM only increased the current density and not the V_{oc}. The thermal activation barrier for carrier transport for the MEH-PPV and P3HT solar cells were consistent with previously measured literature values. These devices were measured to exhibit positive efficiency temperature coefficients. A new passivation technique using PMMA and parylene was employed and showed promise for improving polymer device longevity under terrestrial environmental exposure.

ACKNOWLEDGMENTS

Financial support for this project was made by BP Solar, NASA (NNC056A60G), and AFOSR (SUNY Buffalo, PI Paras Prasad) (FA9550-06-0398).

REFERENCES

[1] K. Kim, J. Liu, M. A. G. Namboothiry, D. L. Carroll, Applied Physics Letters 90 (2007) 163511/163511-163511/163513.

[2] E. C. Chang, C.-I. Chao, R.-H. Lee, Journal of Applied Polymer Science 101 (2006) 1919-1924.

[3] R. C. Hiorns, R. de Bettignies, J. Leroy, S. Bailly, M. Firon, C. Sentein, A. Khoukh, H. Preud'homme, C. Dagron-Lartigau, Advanced Functional Materials 16 (2006) 2263-2273.

[4] F. C. Krebs, J. E. Carle, N. Cruys-Bagger, M. Andersen, M. R. Lilliedal, M. A. Hammond, S. Hvidt, Solar Energy Materials & Solar Cells 86 (2005) 499-516.

[5] F. C. Krebs, Solar Energy Materials & Solar Cells 90 (2006) 3633-3643.

[6] G. Dennler, C. Lungenschmied, H. Neugebauer, N. S. Sariciftci, M. Latreche, G. Czeremuszkin, M. R. Wertheimer, Thin Solid Films 511-512 (2006) 349-353.

[7] W. Ma, J. Y. Kim, K. Lee, A. J. Heeger, Macromolecular Rapid Communications 28 (2007) 1776-1780.

[8] S. Guenes, H. Neugebauer, N. S. Sariciftci, Chemical Reviews (Washington, DC, United States) 107 (2007) 1324-1338.

[9] R. A. e. a. DiLeo, Determination of Nanomaterial Energy Levels for Organic Photovoltaics by Cyclic Voltammetry, in: MRS, Boston, 2007.

[10] S. C. Jain, T. Aernout, A. K. Kapoor, V. Kumar, W. Geens, J. Poortmans, R. Mertens, Synthetic Metals 148 (2005) 245-250.

[11] K. Kim, J. Liu, M. A. G. Namboothiry, D. L. Carroll, Applied Physics Letters 90 (2007) 163511/163511-163511/163513.

Mater. Res. Soc. Symp. Proc. Vol. 1031 © 2008 Materials Research Society   1031-H09-50

Rigid-Rod Sensitizers bound to Semiconductor Nanoparticles

Olena Taratula, and Elena Galoppini
Chemistry Department, Rutgers University, 73 Warren Street, Newark, NJ, 07102

ABSTRACT

A series of "rigid-rod" dyes with an organic chromophore (pyrene or coumarin) attached through an oligophenylenethynylene (OPE) rigid bridge, linear or branched, to an anchoring isopthalic acid unit (Ipa) were synthesized and studied for solar cells (DSSCs) applications. The new dyes were attached to metal oxide (MO_n = TiO_2, ZrO_2 and ZnO) nanoparticles films via the two COOH binding groups on the Ipa unit to investigate their binding and photophysical properties at the semiconductor surface. FTIR-ATR spectra show that all dyes did bind to the metal oxide films through carboxylate bonds. Fluorescence emission on insulating ZrO_2 films was employed to study aggregation of the organic rigid-rods. Studies of the pyrene rigid-rods in solar cells showed near quantitative conversation of absorbed photons into electricity.

INTRODUCTION

In DSSCs, a dye molecule is covalently attached to a nanostructured semiconductor oxide film. Upon light absorption, an electron is transferred from the excited dye to the semiconductor. Then the electron diffuses through the semiconductor nanoparticles to the electrode and the photoelectrochemical cycle is completed by electron donation to the oxidized dye (regeneration) by a redox system in the electrolyte [1]. The molecular design of novel photosensitizing chromophores is an important area of DSSCs research.

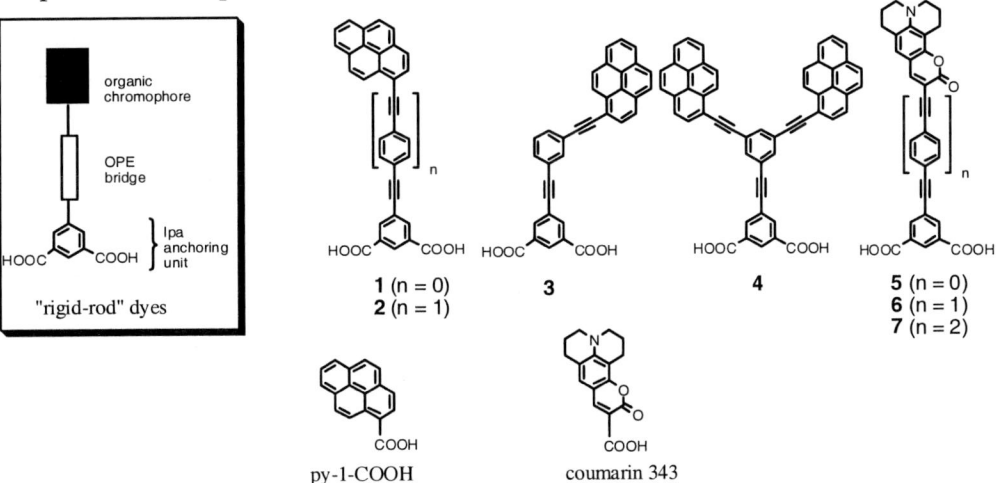

Figure 1. Organic rigid-rods with pyrene (**1-4**) and coumarin (**5-7**) described in this study, and references compounds.

Lately, organic sensitizers anchored through rigid linkers have been developed to study the interfacial electron injection [2]. Such models are used to control the distance between the chromophore and the semiconductor surface and tune the photophysical properties of the dyes, for instance, increasing extinction coefficients by extending the conjugation. In this paper we describe our recent studies of a series of organic rigid-rods (Figure 1). The properties of novel compounds **5-7** were compared to the pyrene series (**1-4**) that we had previously described [3, 4]. In particular, the influence of OPE bridge length and branching on photophysical and binding properties of rigid-rods was investigated.

EXPERIMENT

Pyrene-1-carboxylic acid, coumarin 343 and deoxycholic acid (DCA) were purchased from Aldrich. The synthesis of **1-4** [3, 4] and **5-7** [5] has been described. TiO_2, ZrO_2 and ZnO colloidal nanoparticle thin films were prepared by well-known sol-gel techniques [6, 7]. For binding studies, the colloidal films were cast onto cover glass slides, and fluorine-doped tin oxide (FTO) glass (7 Ω for sq.cm, Pilkington) was used for photoelectrochemical studies. After sintering at 450°C for TiO_2 and ZrO_2, and at 350°C for ZnO, colloidal films with ~ 10 µm thickness were obtained. The binding of **1-7** was accomplished by immersing the films into 0.5 to 1 mM THF, solutions for 30 min or 12 hrs at room temperature. In some cases, prior to binding the films were treated with acid (HCl aq., pH 1) or base (NaOH aq., pH 11). Fourier transform infrared attenuated total reflectance (FTIR-ATR) spectra were recorded on a Thermo Electron Corporation Nicolet 6700 FT-IR. UV-Vis absorption spectra were acquired on a Varian Cary-500, and Fluorescence emission spectra on Varian Cary-Eclipse. The Incident-Photon-to-Current Efficiencies (IPCE) were quantified in a sandwich-cell arrangement with a 0.5 M LiI/0.05 M I_2 acetonitrile electrolyte.

DISCUSSION

UV-Vis spectroscopy

Table I summarizes the UV-Vis data for compounds **1-7** in THF solutions. Among all four studied pyrene rigid-rods, the absorption spectra of **2** exhibited the largest bathochromic shift (Table I), in solution and on the film, due to the presence of the OPE bridge. The long wavelength absorbance band of **3** and **4,** with the PE substituted in *meta* position, was blue shifted due to the broken π-conjugation of OPE linker. Upon binding to the TiO_2 surface, absorption broadened and was red shifted, as shown in the spectra for the absorbance band of **4** (Figure 2a and 2b).

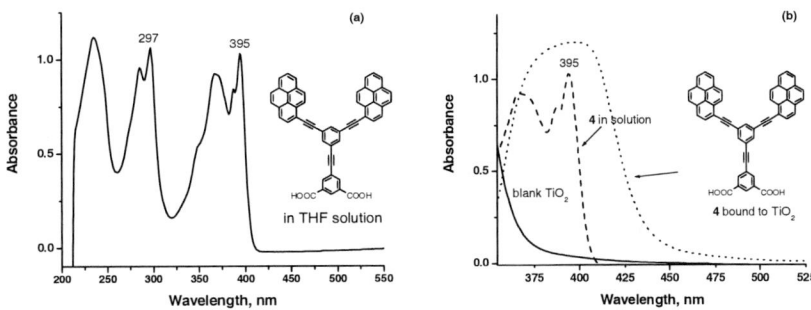

Figure 2. Absorption spectra of **4** *(a)* in THF solution and *(b)* bound to TiO$_2$, overlaid with blank TiO$_2$ and dye in THF solution. In spectrum *b*, the absorption below 375 nm is obscured by the absorption of the semiconductor.

Table I. Photophysical properties of organic dyes **1-7**, and reference compounds, pyrene-1-carboxylic acid and coumarin 343, in THF solutions.

	[a]**Py-1-COOH**	[b]**1**	[b]**2**	[b]**3**	[b]**4**	**Coumarin 343**	**5**	**6**	**7**
λ_{abs}, **nm**	351	384	402	385	395	441	443	448	452
ε, ×10^4 **M^{-1}cm^{-1}**	3.2	3.6	5.6	4.3	7.2	4.0	4.2	4.5	4.8
λ_F, **nm**	390	425	445	417	402, 425	477	486	497	503

[a]From reference 8
[b]From reference 4

The absorbance spectra of coumarin rigid-rods **5-7** showed a broad absorption band in the visible range at ~450 nm (Figure 3a) typical of the coumarin chromophore and assigned to the intramolecular charge-transfer transition from the electron donor (amino-group) to the electron acceptor (carbonyl group). The influence of the OPE length on photophysical properties of coumarin rigid-rods **5-7** is shown in Table I: with the increase of the OPE length the extinction coefficient increases and the ICT band is shifted to longer wavelengths.

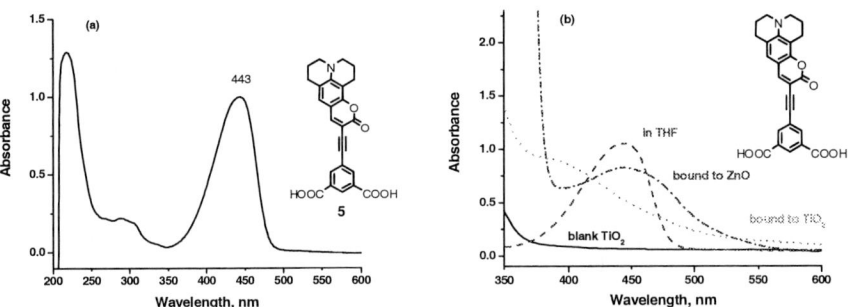

Figure 3. Absorption spectra of **5** *(a)* in THF solution and *(b)* bound to MO$_n$: TiO$_2$ (green line) and ZnO (blue line) compared with blank TiO$_2$ (black line) and the dye is solution (red line). In spectrum *b*, the absorption below 375 nm is obscured by the absorption of the semiconductor.

Upon binding to TiO$_2$, the ICT band of **5-7** was blue shifted (see, for instance, the spectrum of **5**, in Figure 3b) and red shifted upon binding to ZnO (Figure 3b, blue line). Although shifts can be attributed to the formation of aggregates on the surface of the semiconductor, the reason for the differences observed on the two different MO$_n$ substrates is under investigation. Overall, compounds **5-7** were thermally unstable. Degradation of coumarin rigid-rods bound to ZnO was found to be much slower when compared to the same compounds bound to TiO$_2$.

FT-IR-ATR measurements

All rigid-rods were covalently bound to MO$_n$ films via the COOH anchoring groups of the Ipa unit. FTIR-ATR spectroscopy was used to determine the type of chemical bonds that form upon binding to the semiconductor surface. The FTIR-ATR spectra of the pyrene rigid-rods have been reported [4]. The IR spectra of coumarin rigid-rods **5-7** (in Figure 4, the spectrum of **6** is shown as an example) showed C=O stretches at ~1720 and 1695 cm^{-1} due to the carbonyl group of the coumarin ring and the COOH anchoring groups (Figure 4). Upon surface binding, the 1695 cm^{-1}-band disappeared and bands assigned to the carboxylate groups at ~1600 and 1400 cm^{-1} appeared. These spectral changes are consistent with what has been observed for **1-4** and other dyes bound through COOH groups, and are indicative of bidentate bond formation (Figure 4, inset) [2].

In addition, the surface binding of amino-coumarin rigid-rods was found to be dependent on the pH treatment of the metal oxide surface prior to the binding step. Upon binding to pH 11 pretreated TiO$_2$ film, the carboxylate binding was also observed. However, on pH 1 pretreated films only one new stretch at ~ 1630 cm^{-1} was observed (Figure 4, inset). Table II summarizes IR stretching frequencies of **6** upon binding to TiO$_2$ pretreated with acidic or basic aqueous solutions (see experimental details). Similar binding behavior was observed for all pyrene and coumarin rigid-rods (**1-7**). The highest surface coverage of the dye on MO$_n$ surface was achieved using the pH 11 pretreated films whereas the dyes weakly bound to pH 1 pretreated films.

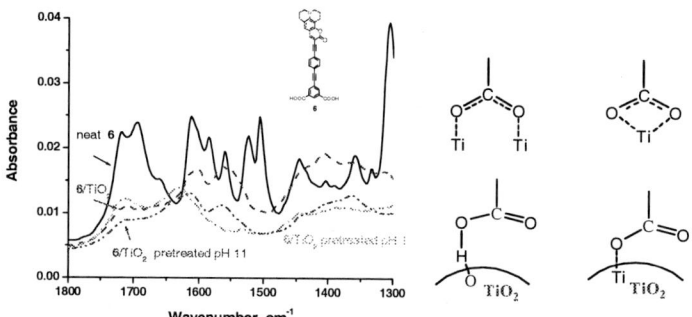

Figure 4. FT-IR-ATR spectra of **6** in the solid state (black line) and bound to TiO$_2$: untreated TiO$_2$ (red line), TiO$_2$ pretreated with acid at pH 1 (green line), and TiO$_2$ pretreated with base at pH 11 (blue line). Inset: Schematic representation of possible binding modes.

Table II. IR stretching frequencies (cm^{-1}) for **6**, neat or bound to TiO$_2$ films untreated and TiO$_2$ pretreated with acid or base.

	C=O	Aromatic C=C	$(COO^-)_{as}$	$(COO^-)_{sym}$
neat 6	1720, 1695	1613	-	-
bound/untreated	1718	1609	1560	1405
bound/pH 1	1713	1608	-	-
bound/pH 11	1715	1612	1565	1365

Fluorescence emission in solution and on ZrO$_2$

Rigid-rod sensitizers **1-7** exhibited room temperature fluorescence emission in THF solution. Fluorescence emission maxima (λ_F in Table I) were found to be dependent on the OPE linker's length and branching. In all cases, fluorescence emission was quenched upon binding to TiO$_2$ suggesting efficient electron injection into the semiconductor. The emission spectra for **1-4** bound to insulating ZrO$_2$ (bandgap = 5 eV) showed excimer emission at ~520 nm (Figure 5a), and no evidence of monomer emission. Organic rods were found to aggregate on the metal oxide surface.

Upon binding of the amino-coumarin rigid rods to ZrO$_2$, almost all fluorescence emission was quenched (Figure 5b) suggesting the formation of non-emissive aggregates. However, when the dye was co-bound with deoxycholic acid (DCA), the aggregation of the coumarin dyes on the film surface was prevented (Figure 5c).

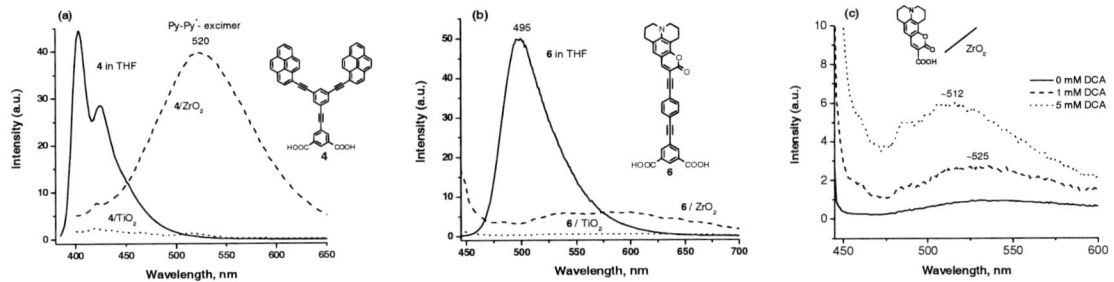

Figure 5. Fluorescence spectra of *(a)* **4** and *(b)* **6** in THF solution (solid line) and bound to TiO$_2$ (dotted line) and ZrO$_2$ (dashed line). *(c)* Fluorescence spectra of coumarin 343, as an example, bound to ZrO$_2$ without (solid line) and co-bound from DCA solutions (1 mM and 5 mM).

Photoelectrochemical studies

IPCE values were reported for pyrene rigid rods [4]. All studied pyrene rigid-rods showed high IPCE value, ~30% for **1** and **3** and ~45% for **2** and **4** with the absorptance (percentage of light absorbed) close to unity. High photocurrent efficiencies were observed at surface coverage at which the excimer formation on ZrO$_2$ was observed. Since pyrene rigid rods showed good efficiency in converting light into electricity at high surface coverages, it is probable that the pyrene excimer behaves as the sensitizer, effectively injecting electrons into semiconductor. The IPCE studies of **5-7** are in progress.

CONCLUSIONS

The photophysical properties of a series of pyrene and coumarin rigid-rod dyes were described and found to be dependent on the length and branching of the OPE linker unit. Each additional PE unit shifts the absorbance of the dye to the red except in the case of meta-substituted PE units. FT-IR-ATR measurements showed that rigid-rods form the carboxylate bonds on the neutral or basic metal oxide surface. The formation of pyrene excimer was observed in the emission spectra of **1-4** bound to insulating ZrO_2. Pyrene rigid rods showed good efficiency in converting light into electricity, suggesting that the pyrene excimer behaves as the sensitizer. The highest surface coverage of the dyes was achieved by binding to pH 11 pretreated TiO_2 films.

ACKNOWLEDGMENTS

The authors are thankful to the National Science Foundation (NIRT-0303829) and to the American Chemical Society (PRF-4663-AC-10) for the funding. We are grateful to Dr. Mendelsohn for generous access to his FT-IR-ATR instrument and Dr. Dong Wang for samples of **1** and **2**.

REFERENCES

1. K. Kalyanasundaram and M. Grätzel, *Coord. Chem. Rev.* **77**, 347 (1998).
2. E. Galoppini, *Coord. Chem. Rev.* **248**, 1283 (2004).
3. D. Wang, J. M. Schlegel, and E. Galoppini, *Tetrahedron* **58**, 6027 (2002).
4. O. Taratula, J. Rochford, P. Piotrowiak, E. Galoppini, R. A. Carlisle and G. J. Meyer, *J. Phys. Chem. B* **110**, 15734 (2006).
5. O. Taratula and E. Galoppini, *J. Phys. Chem. B* Manuscript in preparation.
6. T. A. Heimer, S. T. D'Arcangelis, F. Farzad, J. M. Stipkala and G. J. Meyer, *Inorg. Chem.* **35**, 5319 (1996).
7. O. Taratula, E. Galoppini, D. Wang, D. Chu, Z. Zhang, H. Chen, G. Saraf and Y. Lu *J. Phys. Chem. B* **110**, 6506 (2006).
8. P. G. Hoertz, R. A. Carlisle, G. J. Meyer, D. Wang, P. Piotrowiak and E. Galoppini, *Nano Lett.* **3**, 325 (2003).

Mater. Res. Soc. Symp. Proc. Vol. 1031 © 2008 Materials Research Society   1031-H09-54

Determination of Nanomaterial Energy Levels for Organic Photovoltaics by Cyclic Voltammetry

Roberta Ann DiLeo[1], Annick Anctil[2], Brian Landi[1], Cory Cress[2], and Ryne P Raffaelle[1]
[1]Rochester Institute of Technology, Rochester, NY, 14623
[2]Microsystems Engineering, Rochester Institute of Technology, Rochester, NY, 14623

ABSTRACT

A wide variety of nanomaterials and associated nanomaterial/polymer composites are being developed in an effort to produce higher efficiency organic solar cells. This development requires a fundamental understanding of the energy levels for the individual materials, and their composites, to enable device designs which posess appropriate energy level matching. Cyclic voltammetry (CV) allows for the determination of the band gaps (E_g) and energy levels of these various nanomaterials and composites by measuring their oxidation and reduction potentials. These potentials correspond to a given material's ionization potential (IP) and electron affinity (EA), respectively. The results for the EA, IP, and E_g have been determined by CV for derivatized fullerenes and CdSe quantum dots (QD), measured in isolation, and in conjugated polymer composites with MEH-PPV. In addition, CV measurements conducted under dark and illuminated conditions were used to investigate the relationship between energy levels within the composites.

INTRODUCTION

The rapidly growing field of organic photovoltaics (PV) is pushing the limits of achievable efficiencies with conventional materials [1, 2]. The development of high efficiency devices requires an assembly of compatible materials which support photon absorption, exciton diffusion, exciton dissociation, and carrier transport [3]. Choice of suitable materials and their energy level alignments are critical to the development of efficient devices. The enhancement of photon absorption in a device can be accomplished by selecting materials with appropriate band gaps to absorb the solar spectrum. It is also important to minimize recombination sites within the bulk junction; this can be done by avoiding materials which will cause traps due to their unsuitable energy level alignment. Cyclic voltammetry (CV) is a characterization technique that offers the capabilities of probing these important material properties. It is possible to calculate a material's ionization potential (IP) and electron affinity (EA), which correspond to valence states and conduction states, respectively, using CV.

Electrochemical CV measurements can be used to determine these potentials by measuring the voltages at which the material undergoes reduction or oxidation with respect to a reference electrode. These values can be used to calculate the EA and IP of a material as explained by Kucur et al. [4]:

$$EA = -(E_{red} + \Delta E_{vacuum,electrode}) \quad (1)$$

$$IP = -(E_{ox} + \Delta E_{vacuum,electrode}) \quad (2)$$

The band gap (E_g) can be obtained by taking the energy difference of the EA and IP.

There have been several reports on the cyclic voltammetry of nanomaterials and polymers commonly used in organic photovoltaics [5-8]. In addition, some research groups have used CV to examine the interaction of CdSe QDs and nanomaterials with polymers in a composite [9-11]. A next step in understanding solar cells is the examination of the pure materials and respective composites under illumination, which yields a representative environment for working devices.

This work has focused on determining the energy levels for a series of CdSe quantum dots, various fullerene derivatives, and the conjugated polymer MEH-PPV (Poly[2-methoxy-5-(2-ethylhexyloxy)-1,4-phenylenevinylene]). CV was used to measure composites of these materials under both dark and illuminated conditions.

EXPERIMENT

CdSe QDs were synthesized using a CdO precursor and were extracted with a mixture of methanol and hexanes as previously reported [12]. A series of four QD sizes were synthesized by varying synthesis time. A ligand exchange of the triocytlphosphine (TOPO) capped QDs was achieved with pyridine, using centrifugation at 6000 rpm for 10 minutes, followed by sonication.

Electrochemical measurements were performed in a three neck cell, under argon, with an electrochemical analyzer from CH Instruments. A rotating platinum disk electrode (Pine Instruments) served as the working electrode, a platinum wire as the counter electrode, and a KCl saturated standard calomel electrode (SCE) as the reference electrode. A 0.1 M solution of acetonitrile and tetrabutylammonium hexafluorophosphate (Aldrich) was used as the electrolyte . Voltammograms were obtained at a scan rate of 20 mV/s and rotation speeds between 50-100 rpm. Illumination studies were performed under dark conditions and under illumination, using halogen lamps (peak λ = 630 nm) and a 350 nm UV lamp. Optical absorption measurements were taken with a Perkin Elmer Lambda 900 (300 to 1600 nm). Fluorescence measurements were taken with a JY-Horiba Fluorolog-3 spectrofluorometer (400 to 800 nm). The reduction and oxidation potentials were used to calculate EA and IP values, respectively.

DISCUSSION

Fig. 1 shows a voltammogram for PCBM C_{60} and the conversion of the two potentials to the respective energy level using a SCE ($\Delta E_{vacuum\ level}$ = 4.158 eV). Fig. 2a shows the optical and fluorescence data for the series of CdSe QDs. The four synthesized dot sizes follow the increasing band gap trend as expected; with smaller dots having a larger band gap. A range of 2.30 to 3.31 nm diameter QDs was calculated based upon Peng's work [13]. Fig. 2b shows the changes in band gap versus QD diameter for optical absorption onsets and peaks, photoluminescence peaks, and electrochemical onsets and peaks. The electrochemical peak values correspond closely with the optical measurements; using the EA and IP peaks rather than the onsets yielded more accurate calculation of band gaps (Fig. 2b). The EA onsets do not change significantly with respect to dot diameter in the electrochemical measurements; whereas the IP onsets change with respect to the band gap (Fig. 2c). In contrast, the EA and IP peak values shift symmetrically with increasing band gap.

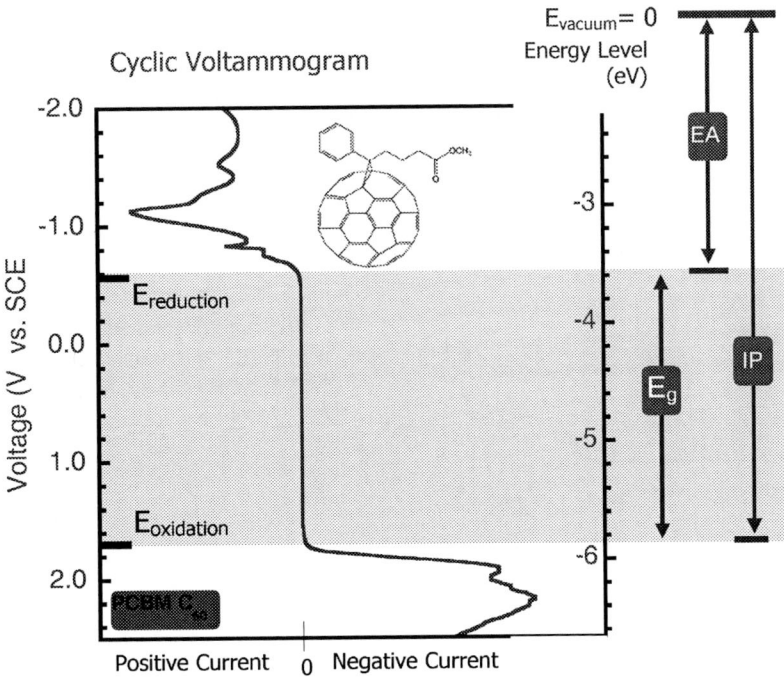

Fig. 1 Representation of a voltammogram of a sample (PCBM C_{60}) and the transition into energy states and band gaps versus vacuum level. (0 V vs. SCE correlates to 4.158 eV with respect to vacuum.)

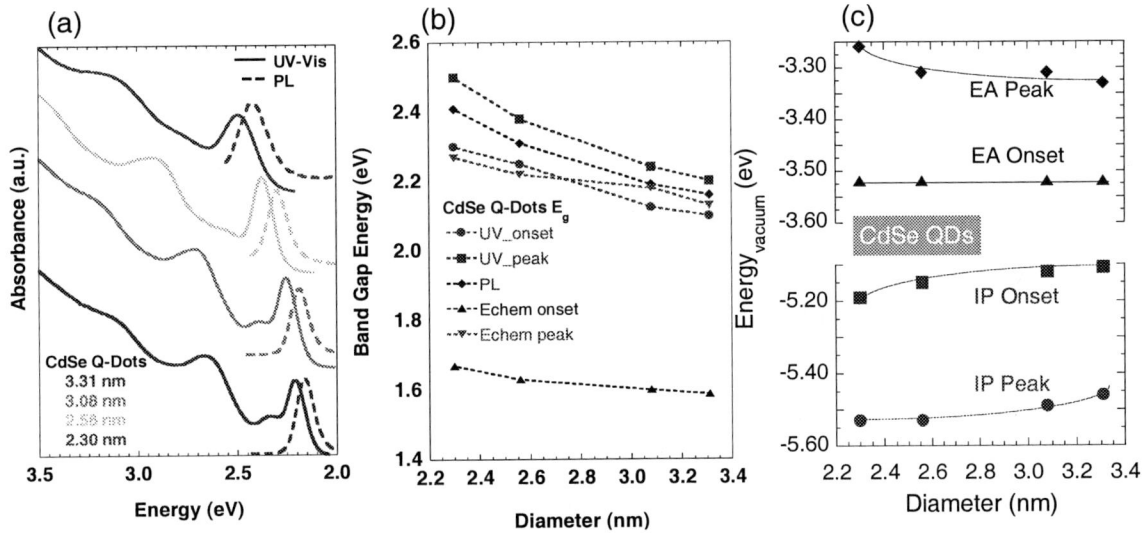

Fig. 2 (a) Optical absorption (solid line) and photoluminescence (dashed line); the QD sizes shown were calculated based upon [13] (b) UV-Vis,, photoluminescence, and electrochemical band gap measurements (c) EA and IP peak and onset values of the QD series. The trend lines in (b) and (c) are guides to the eye.

Fullerenes and their derivatives are commonly used in organic PV devices as electron acceptors. These materials have been successfully employed in a variety of organic PV devices [14-17]. The baseline material of PCBM C_{60} ([6,6]-Phenyl C_{61} butyric acid methyl ester) was characterized along with three derivatives (Fig. 3a-d). PCBM C_{70} ([6,6]-Phenyl C71 butyric acid methyl ester), (Fig. 3b) shows different values for energy states as compared to PCBM C_{60} (Fig. 3a), despite the identical derivatized chains on these two materials. This suggests that fullerenes exhibit different electronic energy states, and in the case of these two derivatives the PCBM C_{70} has a shift in the reduction peak as compared to the PCBM C_{60}.

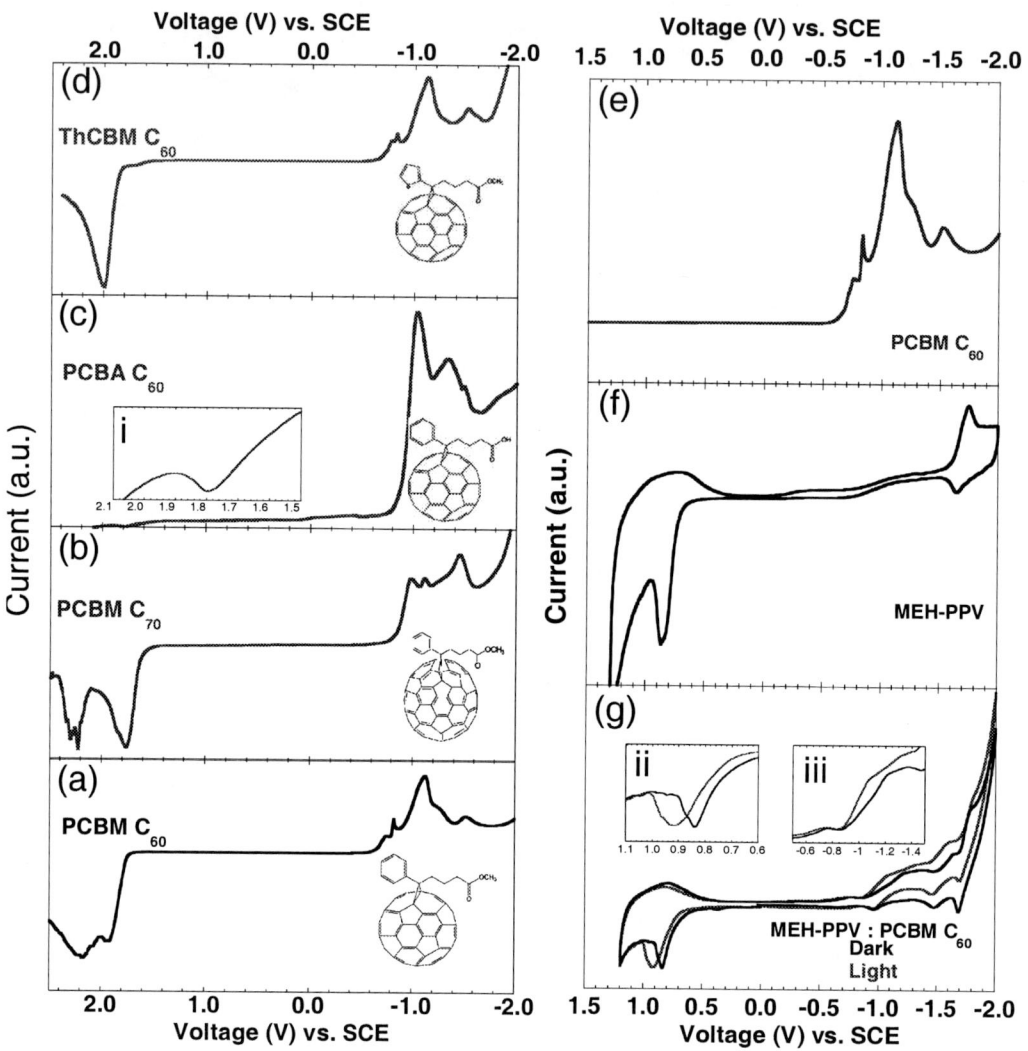

Fig. 3 Voltammagrams of 4 derivatized fullerenes (a) ThCBM C_{60} (i) expanded region of IP (b) PCBA C_{60}, (c) PCBM C_{70}, (d) and PCBM C_{60} (e) PCBM C_{60}, (f) MEH-PPV, (g) MEH-PPV: PCBM C_{60} composite (ii) expanded region of IP (iii) expanded region of of EA

The acid derivative PCBA C_{60} ([6,6]-Phenyl C_{61} butyric acid) (Fig. 3c) also shows a shift in the EA and IP values. This shift in energy levels, compared to PCBM C_{60}, for the acid derivative suggests the chain properties play a significant role in the behavior of the molecule. As a fourth derivative, ThCBM C_{60} (1-(3-methoxycarbonyl) propyl-1-thienyl-[6,6]-methanofullerene) (Fig. 3d) gives rise to behavior similar to PCBM C_{60}, with an EA of -3.50 eV versus the vacuum level, suggesting the substituted ring structure (i.e. phenyl vs. thiophene) has little bearing on EA value for this structure. However, there is an observable change in the IP values upon the thiophene substitution. Fig. 3e and 3f shows the voltammograms of the pure PCBM C_{60} and pure polymer, respectively. The voltammogram of the composite is given in Fig. 3g. The observed energy states in this material are a result of assimilation and shifting of the pure materials' energy states.

Fig. 3g also shows the effects on the MEH-PPV : PCBM C_{60} Composites when the CV is performed under illumination. There is a clear shift to more positive voltage in the IP when the composite is illuminated (Fig. 3gii). A smaller shift in the EA is also observed as shown in Fig. 3giii. This series of CV measurements is summarized in Fig. 4. Although it would not be expected, it is clear from this data that the states observed in the composite are not merely a superposition of the pure materials. The effects on energy levels due to illumination of the composite are also shown red.

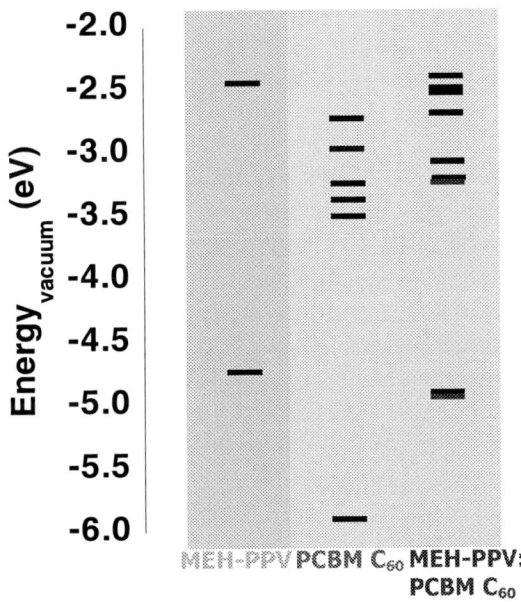

Fig. 4 Energy level cascade of pure PCBM C_{60}, MEH-PPV, and the composite material versus vacuum level. Red bands correspond to shifts due to illumination

CONCLUSIONS

CV was performed on some nanomaterials, polymers and their composites that are commonly used in the fabrication of organic photovoltaics. A series of CdSe QDs with differing diameters was studied. It was shown that for increasing band gap (decreasing dot size) the EA onset remains relatively constant while the IP onset shifts according to band gap. Conversely, the EA and IP peak values shift symmetrically with band gap. The results from the series of fullerene derivatives suggest that both fullerene size and chain composition influence the behavior of energy states. Materials in their pure states were investigated and compared to resulting composites in MEH-PPV. The measurements made on these composites are indicative of energy state behaviors in a device structure. Composite materials were also investigated using controlled illumination environments and shown to exhibit different behaviors under illumination. It was shown for the MEH-PPV: PCBM C_{60} blend that shifts occur in both the EA and IP values, under illumination.

ACKNOWLEDGMENTS

Financial support for this project was made by BP Solar, AFOSR (SUNY Buffalo, PI Paras Prasad) (FA9550-06-0398).

REFERENCES

1. Forrest, S.R., MRS Bulletin, 2005. **30**: p. 28-32.
2. Koster, L.J.A., et al., Physical Review B, 2005. **72**: p. 085205-1-9.
3. Lemaur, V., et al., Journal of American Chemical Society, 2005. **127**: p. 6077-86.
4. Kucur, E., et al., Journal of Chemical Physics, 2003. **119**(4): p. 2333-2337.
5. Zhang, H., et al., Journal of Applied Polymer Science, 2005. **100**: p. 3634-3640.
6. Holt, A.L., J.M. Leger, and S.A. Carter, Journal of Chemical Physics, 2005. **123**: p. 044704-1-7.
7. Kim, J.H. and H. Lee, Chemical Materials, 2002. **14**: p. 2270-2275.
8. Richter, M.M., et al., Chemical Physics Letters, 1994. **226**: p. 115-120.
9. Ding, H., et al., Electrochemistry Communications, 2002. **4**: p. 503-505.
10. Kucur, E., et al., Journal of Chemical Physics, 2004. **120**: p. 1500-1505.
11. Sonar, P., et al., Materials Research Bulletin, 2006. **41**: p. 198-208.
12. Landi, B.J., et al., Materials Letters, 2006. **60**: p. 3502-3506.
13. Yu, W.W., et al., Chemistry of Materials, 2003. **15**: p. 2854-2860.
14. Backer, S.A., et al., Chemical Materials, 2007. **19**: p. 2927-2929.
15. Benson-Smith, J.J., et al., Advanced Functional Materials, 2007. **17**: p. 451-457.
16. Kooistra, F.B., et al., Organic Letters, 2006. **9**: p. 551-554.
17. Zheng, L., et al., Journal of Physical Chemistry B, 2004. **108**: p. 11921-11926.

Synthetic approaches to study aggregation of tripodal linkers on semiconductor surfaces

Sujatha Thyagarajan, and Elena Galoppini
Department of Chemistry, Rutgers University, 73 Warren Street, Newark, NJ, 07102

ABSTRACT

Two tripod-shaped adamantane derivatives carrying a pyrene chromophore and three carboxylic acid binding groups, and varying in footprint size (~ 0.7 and 2.7 nm^2), were synthesized as models to study how the footprint size can influence the aggregation of organic dyes bound to ZrO$_2$ thin films. Synthetic approaches and binding properties of large footprint tripodal linkers are discussed.

INTRODUCTION

Dye-sensitized solar cells (Grätzel solar cells or DSSCs) [1] are developed in research groups worldwide for solar energy conversion. Such devices are based on the sensitization of a semiconductor (usually a mesoporous thin film of TiO$_2$) by a dye covalently bound to the surface of the semiconductor. Inorganic complexes, especially Ru(II) complexes of bipyridines, continue to be preferred because of their stability, but there is an increasing interest in developing and using organic dyes, such as porphyrins [2], coumarins [3] and perylenes [4]. Organic dyes possess useful properties, including the possibility, through careful molecular design, to tune the position of the excited state energy level relative to the conduction band edge of the semiconductor, extend the spectral absorption range, and increase the extinction coefficient. Finally, organic dyes are excellent model systems for fundamental studies of charge injection and recombination processes at the molecule-nanoparticle interfaces. A potential drawback with organic dyes is their tendency to aggregate on the surface of the semiconductor, usually by π stacking interactions. This induces shifts in the absorption spectra, formation of excimers, and quenching of the excited state, leading to decreased performance of solar cells. Hence, there is a need to find strategies to control the spacing of dyes on the surface of semiconductors to study aggregation effects [5].

As part of this interest, we studied in the past, tripodal sensitizers having the structure **A** as shown in Figure 1 [6-8].

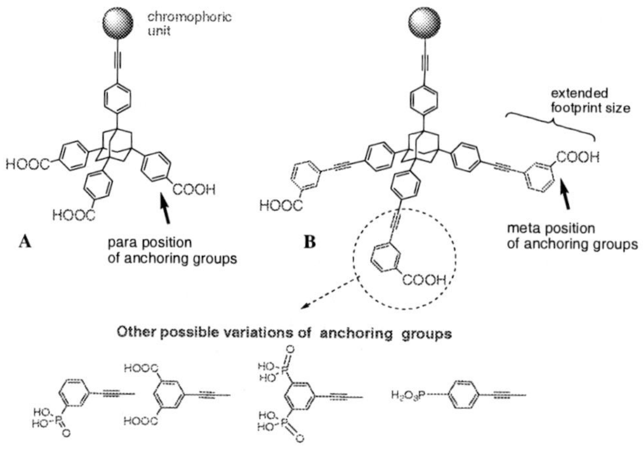

Figure 1. Comparison of tripodal dyes design.

The scope of the tripodal design was to control over the distance and orientation of pyrene with respect to metal oxide surface via the three-point attachment. The studies of tripods having structure as **A**, however, clearly indicated a need to improve this design. First, footprints of variable size are useful to study excimer and monomer effects on the surface of semiconductors. Second, we need a methodology that allows varying the number, type and substitution position of the anchoring functional groups on the tripodal linker to determine the structures that form the strongest binding. This is important, because for most applications, including fundamental charge transfer and excimer effect studies, strong binding and high surface coverages are necessary.

Here we describe the synthesis and properties of tripods having the general structure **B** in Figure 1. Preliminary theoretical calculations indicate that COOH anchoring groups in *meta* position, as in **B**, can improve the binding of compared to the COOH in *para* because of the favorable angle upon anchoring to the metal oxide surface [9, 10]. In addition, this is a more versatile footprint design for tripodal linkers since the anchoring group unit can be changed.

EXPERIMENTAL DETAILS

Synthesis

The synthesis involved the conversion of three the iodo groups in the starting material, tetrakis-1,3,5,7-(*p*-iodophenyl)adamantane (**1**), into COOH groups, only one COOH group in *para* position on each phenyl ring could be introduced, as shown in Scheme 1 [6-8]. The short tripod with pyrene chromophore (**4**) was prepared by the classical Pd-catalyzed Suzuki cross-coupling reaction using the terminal alkyne (**2**) [6] and 1-bromopyrene, followed by hydrolysis. The synthesis of tripodal pyrene chromophore with large footprint size (**7**) in Scheme 2 has been published [10]. Pyrene was selected as the fluorescence probe to make comparisons with the reported pyrene-substituted rigid rods [11, 12] with respect to excimer effects. Pyrene units in close contacts show a characteristic broad structureless band at ~520 nm in the fluorescence emission spectrum, due to excimer formation. The footprint size can prevent aggregation of pyrene and could be useful to study excimer formation.

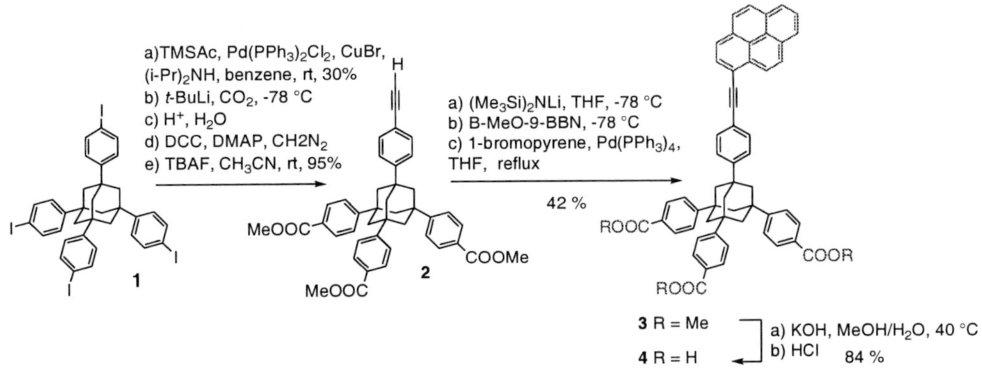

Scheme 1. Synthesis of short tripodal sensitizers.

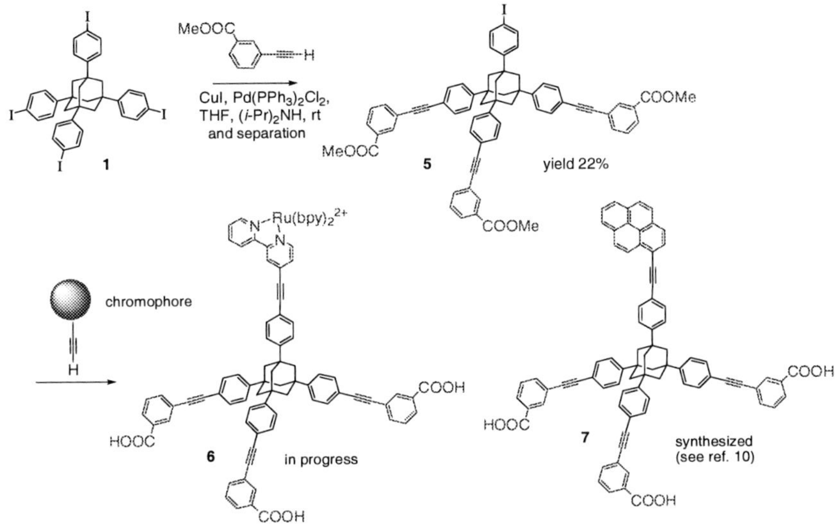

Scheme 2. Synthesis of large footprint tripodal sensitizers.

DISCUSSION

Synthetic strategies

Scheme 3 shows three synthetic strategies were employed to prepare the tripodal sensitizers with varying footprint sizes. The key step in all the routes involves the monosubstitution of the tetrahedral precursor (**I**) to prepare an intermediate having one group different from the other three. The statistical nature of this step requires the separation of di-, tri- and tetra- substituted by-products and therefore demands a lengthy purification procedure. In all routes the chromophoric unit is added in the last step. This method has led to a variety of tripodal sensitizers bearing different chromophores (pyrene, Ru complexes, perylene, and azulene) [6-8].

In Route 1, the p-ethynylenephenylene (PE) bridge is added first to **I** via a Sonogashira cross-coupling reaction leading to a statistical mixture of products. After a separation step, the remaining three iodine groups are converted into *p*-methyl-ester anchoring groups *via* metal-halogen exchange, carboxylation (*t*-BuLi/CO_2), esterification with diazomethane followed by separation using column chromatography to give the alkyne of the tripodal linker (**II**). This order is reversed in Route 2.

In Route 2, the three anchoring groups are introduced first. Cross-coupling of the rigid-rod bridge with the remaining iodophenyl group, followed by TMS deprotection with TBAF gave intermediate **II**. In this approach, since the cross-coupling reaction is non-statistical, the loss of bridging units that are precious (for instance bicyclo[2.2.2]octylene) is avoided [7]. Therefore this route was employed for the synthesis of tripodal linkers with saturated bridges that required a long synthetic procedure and were unexpendable [7].

Route 3 was designed to minimize dimerization of the tripodal alkyne and is the most versatile and shortest approach. The cross-coupling of the rigid-rod bridge containing the *m*-methyl-ester anchoring group to the precursor (**I**), followed by separation gives intermediate **IV**. The advantage of this approach compared to Routes 1 and 2 is that by adding the anchoring unit in the cross-coupling step we are now able to vary the type (PO_3H, $SiOEt_3$, etc.), number (1-5) and position (*o*-, *m*-, *p*-) of the anchoring groups, as well as footprint size. The disadvantage of this approach is that the separation in the first step is particularly difficult.

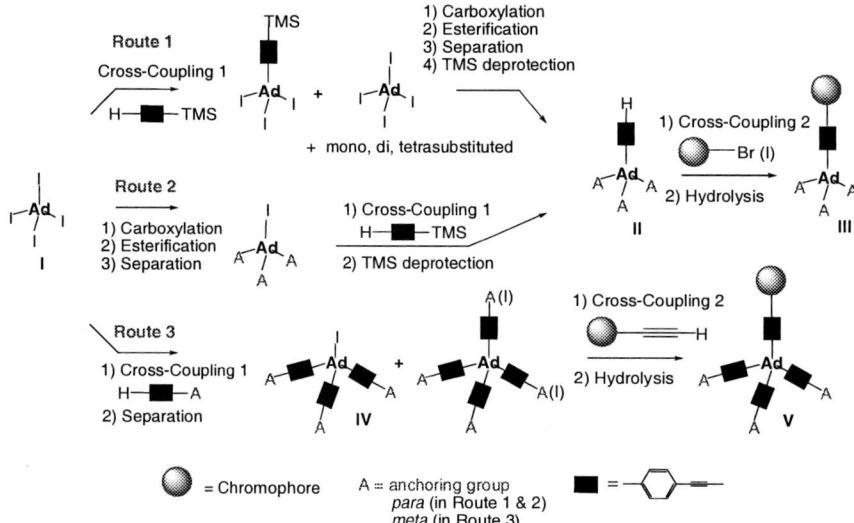

Scheme 3. Outline of synthetic strategies for preparation of tripodal sensitizers.

In general, the cross-coupling reactions proceeded in lower yields as the size of the tripodal linker or of the bridge increased. Detailed synthesis of tripods adopting Routes 1 and 2 [6-8] and Route 3 [10] are published.

Preliminary binding studies for 7

The UV-vis absorption spectra of **7** in THF solution (Figure 2a), displayed spectral changes similar to that observed in other phenylethynylpyrenes [11, 12]. The extension of π-conjugation in **7** resulted in a 20-nm red shift of the longer wavelength band and higher extinction coefficients (ε~36,567 $M^{-1}cm^{-1}$ at 393 nm) when compared to pyrene (ε~200 $M^{-1}cm^{-1}$

at 372 nm) [12]. The bands at 325 nm were assigned to π- π* transitions of the PE units. The UV-vis absorption spectrum of **7** bound to TiO$_2$ is shown in Figure 2b. The higher energy transitions, below 380 nm, are obscured by the absorption of the semiconductor and only the lowest energy vibronic component is observed for **7**.

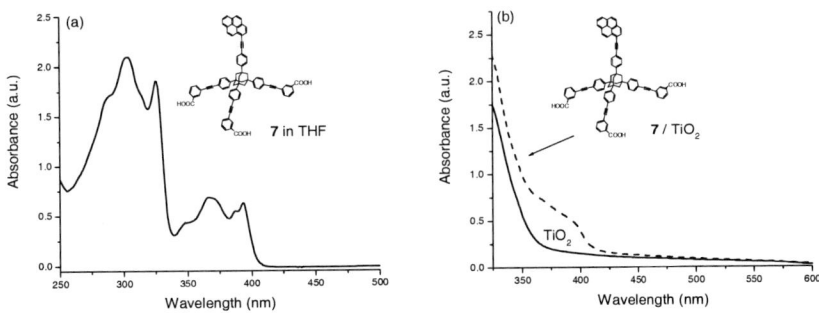

Figure 2. Absorption spectra of **7** (a) in THF solution and (b) bound to TiO$_2$/glass (dashed line) overlaid with TiO$_2$/glass as the reference (solid line).

The fluorescence emission spectrum of **7** in THF solution, Figure 3a, showed that the emission λ$_{max}$ is shifted to longer wavelengths by 25-nm compared with those of pyrene [12]. A quantitative quantum yield (Φ$_{FL}$ ~ 0.9) was obtained for **7** [12]. The pyrene fluorescence of **7** bound to TiO$_2$ was quenched, indicating that the molecules bind to the nanoparticles and that binding results in electron injection in the semiconductor. Emission spectra of **7** bound to ZrO$_2$ (an insulator), however, showed excimer formation at ~ 505 nm, Figure 3b.

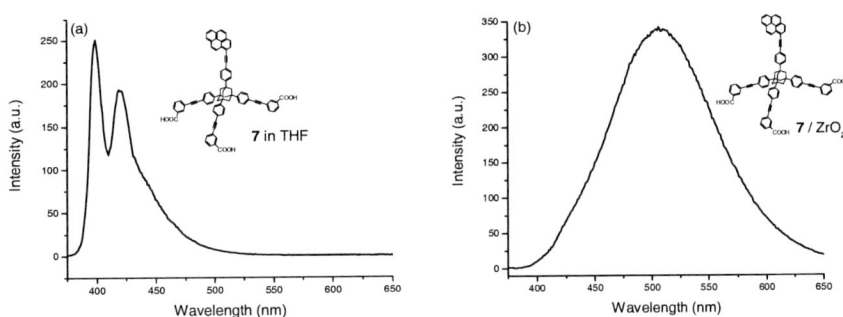

Figure 3. Fluorescence emission of **7** (a) in THF solution, λ$_{ex}$ = 360 nm and (b) bound to ZrO$_2$/glass, λ$_{ex}$ = 360 nm.

Excimer was observed even at low surface coverages, and the footprint size is too large to allow excimer formation. This result indicates that there is extensive contact between pyrene units, probably due to interparticle contacts, as shown in Scheme 4. The thin colloidal films that were employed for the injection studies are made of spherical nanoparticles layers with numerous points of contact and pores. Unbound molecules may remain trapped in the porosities or "touch" nearby particles. This aggregation study is important because it indicates that charge transfer kinetics in this kind of nanoparticle film morphology is complicated by the local inhomogeneity (in addition to other factors such as the dye's excited-state properties).

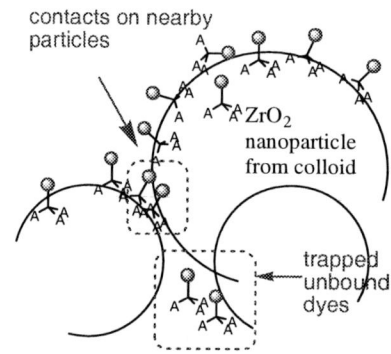

Scheme 4. Interparticle contacts on thin colloidal films of ZrO_2.

CONCLUSIONS

Tripodal linkers with large footprints were prepared to study aggregation and interactions between chromophoric groups on metal oxide surfaces. Binding and excimer studies of these tripods on ZrO_2 films, colloidal diluted solutions and flat surfaces are still in progress, but indicate that excimer is still observed and that the use of large footprints may not be a suitable strategy. The footprint size will also enable us to look into the aspects of charge injection processes like the mechanism of electron transfer into TiO_2.

ACKNOWLEDGMENTS

The authors thank the NSF-NIRT (0303829) and the ACS-PRF (4663-AC10) for research support and Dr. Qian Wei for synthesizing the short pyrene tripod.

REFERENCES

1. K. Kalyanasundaram and M. Grätzel, *Coord. Chem. Rev.* **77**, 347 (1998).
2. W. M. Campbell, A. K. Burrell, D. L. Officer and K. W. Jolley, *Coord. Chem. Rev.* **248**, 1363 (2004).
3. K. Hara, M. Kurashige, Y. Dan-oh, C. Kasada, A. Shinpo, S. Suga, K. Sayama and H. Arakawa, *New J. Chem.* **27**, 783 (2003).
4. L. Wang, R. Ernstorfer, F. Willig and V. May *J. Phys. Chem. B* **109**, 9589 (2005).
5. E. Galoppini and S. Thyagarajan, *The Spectrum* **18**, 22 (2005).
6. W. Guo, E. Galoppini, G. Rydja and G. Pardi, *Tetrahedron Lett.* **41**, 7419 (2000).
7. Q. Wei and E. Galoppini, *Tetrahedron* **60**, 8497 (2004).
8. M. Lamberto, C. Pagba, P. Piotrowiak and E. Galoppini, *Tetrahedron Lett.* **46**, 4895 (2005).
9. P. Persson (private communication).
10. S. Thyagarajan, A. Liu, O. A. Famoyin, M. Lamberto and E. Galoppini, *Tetrahedron* **63**, 7550 (2007).
11. P. G. Hoertz, R. A. Carlisle, G. J. Meyer, D. Wang, P. Piotrowiak and E. Galoppini, *Nano Lett.* **3**, 325 (2003).
12. O. Taratula, J. Rochford, P. Piotrowiak, E. Galoppini, R. A. Carlisle and G. J. Meyer, *J. Phys. Chem. B* **110**, 15734 (2006).

Mater. Res. Soc. Symp. Proc. Vol. 1031 © 2008 Materials Research Society

Energy Band Engineering for Improved Vertical Transport in Quantum Structured III-V p-i-n Solar Cells

Andenet Alemu, and Alex Freundlich
Photovoltaics and Nanostructures Laboratories, Center for Advanced Materials, University of Houston, 724 Science & Research Building 1, Houston, TX, 77204-5002

ABSTRACT

The use of InGaAsN and GaBiAsN quantum structures in the intrinsic region of conventional III-V p-i-n solar cells, lattice matched to GaAs, presents several advantages for photovoltaic (PV) application. First they allow for very shallow to zero valence band offsets thus permitting the free movement of holes. Second, a wide range of band-gap values are made possible due to the large band gap decrease upon the introduction of minute amounts of N and Bi. Using band structure calculations that include the strain effects, conduction and valence band anti-crossing models describing the large band gap bowing and the transfer matrix method, we present the theoretical investigation of optimum design conditions for enhanced vertical transport. The direct quantum mechanical resonant tunneling of electrons out of the quantum structures and into the continuum of the conduction band of the host semiconductor material can be facilitated provided that an adequate choice of material parameters is made. The high electron transmission probability together with the free movement of quasi-3 D holes is predicted to result in enhanced PV device performance. Furthermore, the increase in electron effective mass due to the incorporation of N translates in enhanced absorptive properties, ideal for PV application.

INTRODUCTION

Several inspiring theoretical projections of photovoltaic conversion efficiencies have been made in the past [1-4]. Although these mainly considered idealized materials, their practical realizations have been something of a driving force behind research efforts to identify the right material system. Single band-gap conventional semiconductor solar cells have been the firsts to be extensively studied and developed although their conversion efficiency is limited due to the uniqueness of their cutoff energy [5]. Multi-color, or series connected multiple band-gap solar cells have also seen their fair share of investigation and development. Despite their inherent requirement for current matching and lattice matching between different sub cells, they remain the most powerful PV device developed so far with even higher efficiencies obtained under concentration. Quantum confined solar cells have been proposed as early as 1991 [6]. Despite numerous research efforts on all variations of the concept (quantum wells/wires/dots), their performance still remains less than expected. The main issue in this case is related to recombination losses occurring before carrier escape and collection. Quantum confinement adds energy barriers that hamper the vertical transport, in effect, often hindering the swift transfer of carriers to the electrodes. The comparative advantage of extending the absorption of a solar cell towards the infra-red, through the inclusion of quantum confined structures, may then be completely suppressed due to a very sluggish escape process.

In order to avoid carrier loss, the structures have to be designed in a way that optimizes carrier escape and collection. At operation, thermal escape, is not enough to extract carriers from well potentials that are much deeper than the available thermal energy (~ 25 meV). The use of resonant tunneling of carriers between quantum wells was first proposed in the original paper on quantum well solar cells [6]. The main problem with this method is that it cannot be optimized simultaneously for both electrons and holes. In both the original paper as well as in the later work of O. Raisky et al. [7], the tunneling structures were only optimized for electrons while thermal energy was thought to be enough for escaping holes. Depending on the magnitude of the valence band (VB) offset between the well and the barrier the improvement seen for electron transfer might not simultaneously occur for holes. In order to improve the vertical transport of all carriers we need to look into material systems that display a negligible band offset at either the conduction band (CB) or VB. This will allow us to decouple the electron and hole escape problem and use the resonant tunneling design for the other band. Also, in the absence of a GaAs compatible 1-1.3 eV band gap material the use of quantum structures comes in handy in boosting the photovoltaic conversion efficiency to a new high [8]. We present here band alignment and carrier escape investigation of InGaAsN/GaAs and GaBiAsN/GaAs quantum structures embedded in the intrinsic region of conventional GaAs p-i-n solar cell.

THEORETICAL FRAMEWORK

The observation of unusually large band gap lowering upon incorporation of a small amount of nitrogen in GaAs [9] has sparked interest in the development of dilute nitrogen containing III-V semiconductors for optoelectronic applications. This finding has opened a window on a whole new family of materials that present new opportunities for innovative devices, including photovoltaics [10-13]. From previous experimental studies, an increase in the electronic effective mass by about three fold has been deduced for N mole fraction of 1.9% in GaAsN epilayers grown on GaAs [11]. The subsequent increase of the absorption coefficient (by a factor equivalent to the square root of three ~ 1.73 at the band edge) means that less material will be needed for the absorption of the incoming photon flux. As a consequence, a higher built-in electric field can be maintained in the active region of a quantum-structure-encompassing p-i-n solar cell where such dilute nitride layers form the well material. The increase of the effective mass also allows deeper confinement energies with the same thickness of well material. Very encouraging observations such as a high open circuit voltage and relatively high quantum well photocurrent have also been made on GaAsN MQW structure included in the i-region of a p-i-n GaAs solar cell [13]. A few percent of nitrogen is sufficient to obtain a band gap of desired value by inducing a large decrease of the CB level (~1.1 eV at 1.7% of N). From analysis of band offset in GaNAs/GaAs by X-ray photoelectron spectroscopy, Kitatani et al [14] have reported energy discontinuities in the VB and CB of $-(0.019\pm0.053)$ eV/%N and $-(0.175\pm0.053)$ eV/%N, respectively. Although nitrogen incorporation does not directly alter the VB of GaAs, it have an impact on it via the introduction of strain. The strain induced VB offset between the GaAsN layers and the GaAs could be minimized by incorporating an appropriate element (In, Bi etcetera) in the nitrogen containing material in order to minimize the offset.

Valence band alignment for quasi-3D holes

The valence-band structures were developed within the framework of the $k.p$ formalism including strain and spin orbit interaction. In the case of GaBiAsN the valence band anticrossing model developed by Alberi et. al [15] for GaBiAs was used with modifications for the strain effects induced by the N incorporation. Unlike GaN, GaBi and InAs have a larger lattice parameter than GaAs. The incorporation of the right amount of indium or bismuth in the GaAsN matrix is used to minimize the VB offset. In particular, the similarly large band gap lowering upon incorporation of a small amount of bismuth in GaAs is used to tailor the VB offset independently from the CB. Any strain-induced movement of the GaAsN VB due to the addition of nitrogen can thus be compensated by the incorporation of an appropriate amount of indium or bismuth. In the case of GaBiAsN, the addition of bismuth in GaAs leads to a strong red shift [16, 17] through the formation of a bismuth band that couples with the VB of the host material [15]. The CB of GaAs is left unchanged by the incorporation of bismuth, except for strain effects. The large lowering of the band gap is due to the reduction of the CB minima, through the nitrogen effect, and the increase of the VB maxima through the bismuth effect as described by the conduction [18] and valence [15] band anticrossing models, respectively.

Conduction band resonant tunneling alignment

The band structures were calculated by including the band anticrossing effects on both CB and VB mentioned above, spin orbit couplings and the strain effects. The band structures at the Γ point of the III-V heterostructures were calculated using published band offset ratios. Nevertheless the strain parameters of GaAsN were used instead of those of GaBiAsN in the absence of reliable data for GaBi. The quantum mechanical treatment of the problem in the presence of electric field consists of the solving of Schrodingers' equation of the movement of the electron. We used these solutions combined with the transfer matrix formalism to quantitatively describe the effect of resonant tunneling under electric field. The carrier transmission (tunneling) coefficient is determined using this method [10]. The effect of the electric field was modeled by considering atom thick layers of the i-region materials with a constant potential in each monolayer that has a step like variation between monolayers. This assures the accuracy and validity of the transfer matrix formalism for this system. The right combinations of parameters (i.e composition, thickness etc.) are sought for maximum resonant tunneling under normal operating conditions. Since confinement occurs along the growth direction (z), the movement of the electrons was considered in that direction under the envelope function formalism.

RESULTS AND DISCUSSION

For both InGaAsN (Fig. 1) and GaBiAsN (Fig. 2), the band offsets for both heavy and light holes can be tuned so as to have very small effective barriers, smaller than or equivalent to the available thermal energy (<25 meV). The result will be 3-D or quasi 3-D holes that can move freely, as in the bulk.

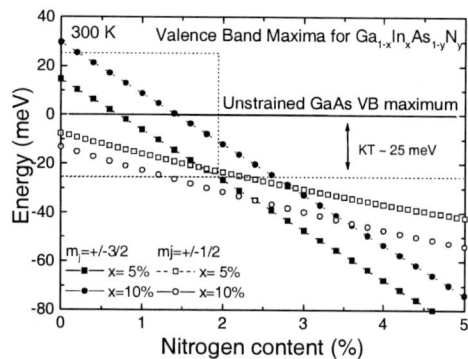

Figure 1: Theoretical calculations of VB offsets between GaInAsN and unstrained GaAs for 5 and 10% indium concentrations vs. nitrogen content up to 5%. Solid symbols are for heavy holes and open symbols for light holes. The solid line represents unstrained GaAs. The big dotted rectangle with a height of ~ 2x25 meV delimits here the In and N mole fractions of interest.

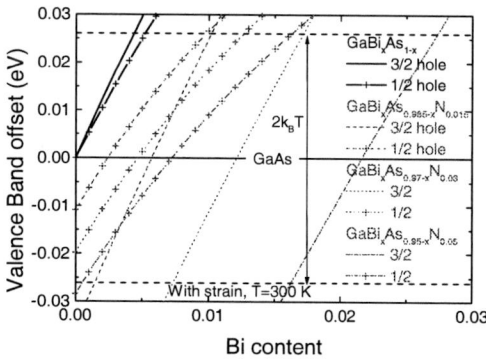

Figure 2: Theoretical calculations of VB offsets between GaBiAsN and unstrained GaAs for 0, 1.5, 3 and 5% of nitrogen concentrations vs. bismuth content up to 3%. Lines without (+) are for 3/2 holes and those with (+) symbol are for ½ holes. The solid line represents unstrained GaAs. The intersection between the solid and open symbols gives the point of lattice matching. Note that the dashed horizontal lines show the +/- k_BT limit delimiting the mole fractions of interest.

As for the CB, our preliminary modeling results show that the value of the electric field needed to align confinement energies between consecutive wells is much closer to the value of the built-in electric field than the field at operation. Fortunately, the weakness of the field at operation also means a much slower energy variation due to the field along the growth direction. It is well-known that a succession of identical wells with small enough barriers results in energy miniband formation in which electrons move freely. It is also known that the application of a strong enough electric field along the coupling direction of the wells breaks the minibands into isolated confinement levels. But for a small enough field, there can still be quantum mechanical tunneling among successive wells for a limited range in the growth direction. Instead of electrically aligning individual wells, we propose to align a set of super lattices with thickness smaller than that range. This results in staircase like energy minibands in the absence of electric field that resonantly couple under the application of the appropriate electric field.

For better illustrative purposes, let's consider $GaBi_{0.01}As_{0.0975}N_{0.015}$ as a good candidate for further examination below. The selection not only satisfies the condition of shallow confinement for holes but is also almost lattice matched to GaAs (according to figure 2), thereby minimizing any strain induced defects in addition to displaying a fairly low band gap of 1.12 eV. The illustration considered here consists of a 50 period $GaBi_{0.01}As_{0.975}N_{0.015}$/GaAs structure. It is partitioned into four different groups with the first three sets of 5, 4 and 3 ML wells while the last is made of 14, 2 ML thick, wells. The thickness of the barriers is kept unchanged at 3 nm. The transmission probability results for various applied electric field are given in figure 5. At zero electric field, the structure presents a high transmission probability only in a narrow energy range close to the band gap of the barrier (1.42 eV). Nevertheless the application of few kV/cm field substantially enlarges this high transmission probability energy range due to increased resonant tunneling effects. This structure is well suited for inclusion in the intrinsic region of a p-i-n GaAs solar cell. It is expected to result in an efficient carrier escape and collection of photo-generated carriers. Several such structures can be included in order to boost the overall absorption and performance of the nanostructured device.

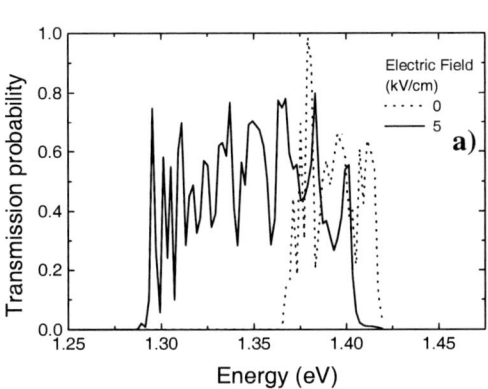 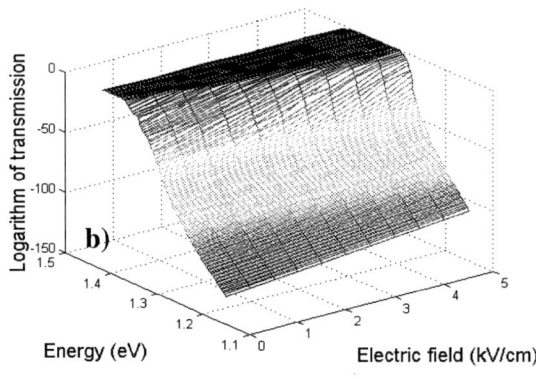

Figure 5: 2D (a) and 3D (b) illustrations of the calculated electron tunneling current transmission coefficient for GaBi$_{0.01}$As$_{0.975}$N$_{0.015}$/GaAs structures with a set of four 12x5 ML, 12x4 ML, 12x3 ML and 14x2 ML wells. The GaAs barriers thickness is 3 nm. The whole structure is lattice matched to GaAs. (a) shows results at 0 and 5 kV/cm while (b) shows the full range of results in 3D. Note the extended range of the maximum of the transmission probability as the value of the electric field increases.

Using an adequate set of material and design parameters, high electron tunneling transmission was achieved for electric field values comparable to the one at operation. The electric field in effect aligns the four sets of minibands thereby facilitating the transfer of electrons from deeper to shallower wells while the barrier height is reduced along the way. This will also lead to gradual increase in carrier escape rate due to greater thermionic escape in shallower wells. However, an efficient tunneling entails that the tunneling escape time [19, 20] is smaller than the radiative and non-radiative recombination lifetimes. Our recent study showed the profound effect that carrier escape dynamics have on nanostructured device performance [21]. The tunneling times corresponding in the above structure are estimated to range from few femtoseconds to few picoseconds thereby satisfying the conditions for fast escape times. Under electric field, the faster spatial separation between tunneling electrons and quasi-3D holes is also expected to help reduce radiative recombinations.

CONCLUSIONS

Promising nanostructured device concepts with staggering theoretical efficiencies where quantum confined states are embedded in the intrinsic region of conventional p-i-n solar cells have been proposed. However practical realizations remain inefficient as these devices suffer from an inherent difficulty in the extraction of photo-generated carriers from the confined states. Within the framework of a "single particle in the box" theory, such shortcomings could be addressed by the use of resonant quantum tunneling designs that can expedite carrier escape. An appropriate choice of material and design is shown here to result in improved vertical transport in quantum structured p-i-n solar cells through the implementation of band alignment and resonant tunneling engineering techniques.

REFERENCES

1. Antonio Luque and Antonio Martí, Phys. Rev. Lett. 78, 5014 - 5017 (1997)
2. Jürgen H. Werner, Sabine Kolodinski, and Hans J. Queisser, Phys. Rev. Lett. 72, 3851 - 3854 (1994)
3. Martin A Green, Nanotechnology 11, 401 - 405 (2000)
4. R. W. Peng, M. Mazzer, and K. W. J. Barnham, Appl. Phys. Lett. 83, 770 – 772 (2003)
5. William Shockley and Hans J. Queisser, J. Appl. Phys. 32, 510 – 519 (1961)
6. K. W. J. Barnham and G. Duggan, J. Appl. Phys. 67, 3490 3493 (1990)
7. O. Y. Raisky, W. B. Wang, and R. R. Alfano, C. L. Reynolds, Jr., D. V. Stampone, and M. W. Focht, Appl. Phys. Lett., Vol. 74, No. 1, 4 January 1999
8. A. Freundlich and A. Alemu, Multi Quantum Well Multijunction Solar Cells for Space Applications, Phys.Stat. Sol. (c), 2 (8), pp 2978-2981 (2005)
9. Markus Weyers and Michio Sato, Appl. Phys. Lett. 62 (12) pp.1396 (1993)
10. L Bhusal, A. Alemu and A Freundlich, Nanotechnology 15 (2004) S245-S249
11. J.A.H. Coaquira, L. bhusal, W. Zhu, A. Fotkatzikis, M.-A. Pinault, A. P. Litvinchuk and A. Freundlich, proceedings of the Fall MRS conference (2004)
12. M. A. Pinault, A. Freundlich, J. A. H. Coaquira, and A. Fotkatzikis, J. Appl. Phys. 98, 023522 (2005)
13. A. Freundlich, A. Fotkatzikis, L. Bhusal, L. Williams, A. Alemu, W. Zhu, J.A.H. Coaquira, A. Feltrin and G. Radhakrishnan, J. Cryst. Growth, 301-302, pp 993-996 (2007)
14. Takeshi Kitatani, Masahiko Kondow, Takeshi Kikawa, Yoshiaki Yazawa, Makoto Okai, and Kazuhisa Uomi, Jap. J. Appl. Phys. 38, 9A (1999) pp. 5003
15. K. Alberi, J. Wu, W. Walukiewicz, K. M. Yu, O. D. Dubon, S. P. Watkins, C. X. Wang, X. Liu, Y.-J. Cho, and J. Furdyna, Phys. Rev. B 75, 045203 (2007)
16. S. Francoeur, M.-J. Seong, A. Mascarenhas, S. Tixier, M. Adamcyk, and T. Tiedje, Appl. Phys. Lett., Vol. 82 (22), 3874-3876 (2003)
17. Wei Huanga, Kunishige Oe, Gan Feng and Masahiro Yoshimotob, J. Appl. Phys. 98, 053505 (2005)
18. W. Shan, W. Walukiewicz, and J. W. Ager, III, Phys. Rev. Lett. 82, 1221-1224 (1999)
19. O. Y. Raisky, W. B. Wang, R. R. Alfano, C. L. Reynolds, Jr. and V. Swaminathana, J. Appl. Phys. 81 (1), 1 January 1997
20. M. Fox, D. A. B. Miller, G. Livescu, J. E. Cunningham and W. Y. Jan, IEEE J. Quantum Electron., vol. 27, pp. 2281-2295, 1991
21. A. Alemu, J. A. H. Coaquira and A. Freundlich, J. Appl. Phys. 99, 084506 (2006)

Evidence of Sequential Carrier Escape in III-V p-i-n Multi-Quantum Well Solar Cells

Andenet Alemu, Jose A. H. Coaquira, and Alex Freundlich
Photovoltaics and Nanostructures Laboratories, Center for Advanced Materials, University of Houston, 724 Science & Research Building 1, Houston, TX, 77204-5002

ABSTRACT

Several InAsP/InP p-i-n Multi-Quantum Well (MQW) solar cells, only differing by their MQW region composition and geometry, were investigated. For each sample, the Arrhenius plot of the temperature related variation of the photoluminescence intensity was used to deduce the radiative recombination activation energy. The electron and holes confinement energy levels in the quantum wells and the associated effective potential barriers seen by each carrier were theoretically calculated. Carrier escape times were also estimated for each carrier. The fastest escaping carrier is found to display an effective potential energy barrier equal to the experimentally determined photoluminescence activation energy. This not only shows that the temperature related radiative recombination extinction process is driven by the carrier escape mechanism but also that the carriers escape process is sequential. Moreover, a discrepancy in device performance is directly correlated to the nature of the fastest escaping carrier.

INTRODUCTION

Recent advances in nanotechnology have invigorated efforts in overcoming conversion efficiency limitations of conventional single junction solar cells [1,2]. In this race for higher efficiencies researchers have proposed various promising device designs incorporating nano-scale components. Nevertheless, to this date, none of the proposed methods have shown any substantial efficiency improvement as theoretically predicted. In fact, most devices display performances far below their conventional counterparts. Multiquantum well solar cells have been proposed as far back as the early nineties [3]. Much discussion has followed on their possible use for solar cell efficiency improvement. By extending the absorption threshold of the device towards the infrared, quantum wells come in handy in increasing the cells photocurrent output. This feature has been shown as being extremely attractive to facilitate current matching in multijunction solar cells and hence to improve the efficiency and radiation tolerance of these devices [4, 5]. Despite obvious advantages for multijunction applications the inclusion of quantum wells is inevitably associated with some amount of open circuit voltage (Voc) degradation. In general, confined states in nanostructured solar cells could act as recombination centers when carriers are not rapidly extracted. Thus, an efficient escape and collection of photo-generated carriers from the well potentials is of critical importance and a necessary path toward higher efficiency devices.

In p-i-n type solar cells, such a condition is satisfied when a large built-in potential exists across the nanostructure containing iregion of the cell [6]. This imposes an upper limit on the thickness of the i-region. Experimentally, best efficiency trade-off is often achieved in the vicinity of this critical thickness where the Voc degradation remains minimal and where a higher photocurrent is afforded by the larger number of wells [6]. But, even for devices that satisfy this condition, occasionally, a severe Voc degradation occurs. To understand these experimental observations and to identify design strategies to prevent I_V characteristics degradation, in this work, we have undertaken a detailed analysis of carrier escape mechanisms.

EXPERIMENT

A set of nearly identical 10-period $InAs_xP_{1-x}$/InP quantum wells solar were fabricated by CBE. Devices had nearly identical intrinsic/ quantum confined regions yielding identical "zero bias" built in electric fields of ~ 30 kV/cm, on the order of the critical one [6]. By setting different arsenic compositions and well thicknesses the quantum confined regions of these devices were engineered to have a similar confined energy gaps but exhibited different electron and hole confined-potential depths. Details concerning fabrication and characterization of p-i(MQW)-n heterostructures have been reported elsewhere [7]. All devices were mesa etched and processed using established recipes. No anti-reflect coating was implemented. Samples studied here can be classified in two categories depending on the value of their Voc (Table 1).

CBE sample No	Ee-hh (eV)	W (Å)	b (Å)	$F_{Built-in}$ (kV/cm)	Voc (V)
672	*1.05*	*18*	*286*	*33.1*	*0.71*
711	*1.0*	*12*	*288*	*33.1*	*0.7*
710	1.0	50	306	29.8	0.6
716	1.0	31	273	33.8	0.62
718	0.9	20	304	31.8	0.62

Table 1. 300K Confined bandgaps (Ee-hh), well (W) and barrier (b) widths, built-in electric field in short circuit conditions (FBuilt-in) and open circuit voltage (Voc) of solar cells studied here. The italicized data corresponds to devices with higher Voc.

THEORETICAL FRAMEWORK

The electron, heavy hole and light-hole energy levels of strained bulk $InAs_xP_{1-x}$ on InP including strain and spin-orbit interaction were calculated. Conduction and valence band offsets were deduced and used in conjunction with well and barrier thickness to determine the potential profile of the $InAs_xP_{1-x}$/InP MQW structures [8]. The solution of the related time independent Schrödinger equation was obtained by the imposing the boundary conditions. The confinement energies of the electron, heavy hole and light hole were determined using a transfer matrix. An example of a typical band alignment diagram (sample 718) is provided in Fig 1.

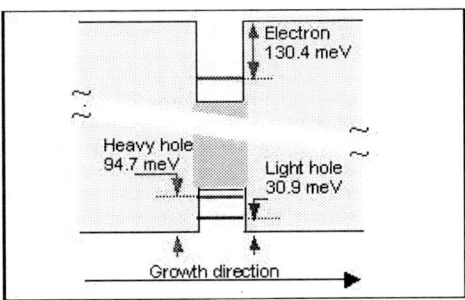

Fig. 1. Schematic representation of quantum well energy band diagram.

Once confinement energies and potential barrier heights for electrons and holes were determined, carrier escape times were extracted using set of equations reported elsewhere [9].

RESULTS

At room temperature Quantum tunneling with characteristic times in the order of femtosecond (10^{-15} sec) dominated the escape of light holes. Thermoionic escape with typical characteristic times in the order of picoseconds (10^{-12} s) prevailed for the escape of heavy holes and electrons (Table. 2). Interestingly for devices having higher open circuit voltages (Voc>0.7 V), the escape time for electrons (determined near operating conditions) was found to be shorter than the one for the heavy holes (Table 2).

CBE sample No	Voc (V)	t_E (10^{-12}s)	t_{HH} (10^{-12}s)	t_{LH} (10^{-12}s)
672	0.71	0.23	0.42	0.09
711	0.7	0.5	0.96	0.08
710	0.6	3.3	1.5	0.4
716	0.62	1.1	0.97	0.2
718	0.62	2.5	2	0.17

Table 2. Open circuit voltage (Voc) and calculated 300 K thermionic escape time for electrons (t_E), heavy holes (t_{HH}) and light holes (t_{LH}) of InP/InAsP MQW p-i-n solar cells.

Although the built-in electric field values of the samples here are similar (~ 32 +/- 2 kV/cm), the devices can be classified in two categories depending on the value of their Voc. The same classification can be obtained when we consider the escape sequence between heavy holes and electrons, as shown in Table 2. The Voc degradation seems to be correlated to the escape sequence. In fact, recombination and escape are two competing processes whereby one increases at the expense of the other [10,11]. Assuming a Voc degradation that is dominated by radiative recombinations [10] one then would expect the activation energy of the luminescence of MQW to provide information on the nature of the carrier that drive the recombination rate. In order to verify this assumption and identify the origin of the open circuit voltage degradation in the samples with lower Voc, photoluminescence (PL) versus temperature measurements were undertaken (Fig. 2-a). As temperature increases to reach 300 K, we expect that the subsequent increase in escape from the quantum wells will be accompanied by a proportional decrease in radiative recombination. The fastest escaping carrier will be driving the escape process. In other words, the fastest escaping carrier will be limiting the radiative recombination rate. By monitoring the quantum wells radiative recombination intensity variation with temperature one can extract the associated activation energy. Our hypothesis is that the value of the effective potential barrier seen by the fastest carrier that thermally escapes from the well will set the value of the PL activation energy. The experimentally obtained activation energy can then be compared to the calculated potential barrier heights in order to verify and determine the carrier escape sequence predicted from theoretical calculations.

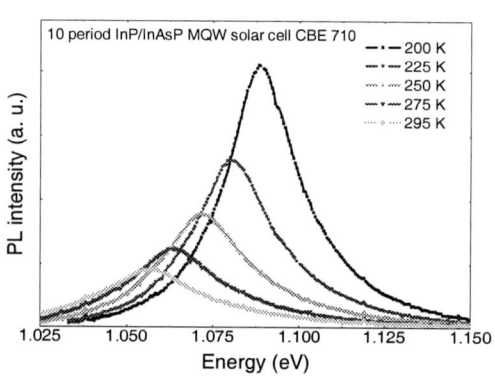
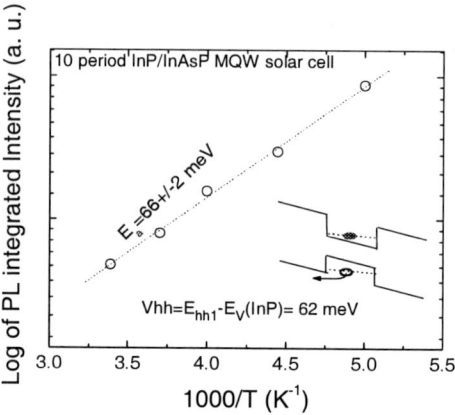

Fig. 2. (a) Temperature dependant photoluminescence of a 10 period InP/InAsP MQW solar cell and (b) Arrhenius plot of the integrated PL intensity (o) as a function of temperature for the InP/InAsP MQW sample CBE 710 with a Voc = 0.6 V. The dotted line is a semi-log fit whose slope is proportional to the activation energy.

PL spectra were recorded for temperatures ranging from 200 K to 300 K where the thermionic escape is dominant for heavy holes and electrons. The radiative recombination activation energy for the well material was extracted from the slope of the Arrhenius plot of the PL intensity as a function of temperature. The PL intensity I(T) , at a given temperature T (in Kelvin), can be expressed by:

$$I(T) = \frac{I_o}{1 + Ae^{-\frac{E_a}{k_B T}}}$$

(1)

where I_0 represents the intensity at 0 K, Ea is the activation energy and A is the amplitude. A semi-log fit can be used for the high temperature side of the curve described by equation (1). The determination of the radiative recombination activation energy of sample CBE 710 is illustrated in Fig. 2-b. The PL activation energy of 66 +/- 2 meV seems to be very close to the calculated heavy hole's potential barrier height of 62 meV (109 meV for electrons). This seems to be consistent with the theoretical prediction that the heavy hole escapes faster than the electron for this sample and, therefore, limits the radiative recombination in the wells. These PL measurements were taken at Voc conditions. Therefore, no current is extracted from the cell and the reduction in radiative recombination in the wells is expected to lead to increased non-radiative recombinations and possibly some additional radiative recombination in the barrier. Unfortunately we do not have a direct measure of the non-radiative process. As for the radiative recombination in the barrier, it too naturally decreases with temperature. A time resolved-spectroscopic technique would be needed to identify any relative increase in such transition due to the decrease of the raditaive recombination in the wells, but this is beyond the scope of this work. Similarly, we have found a very good correlation between the experimentally determined activation energies and the calculated effective potential barrier of the fastest escaping carrier in all our samples. In Table 3, PL activation energies, the calculated effective potential barrier heights seen by both the electron and heavy hole in the well material, together with the values of the Voc are reported. For devices with lower Voc the PL activation energy coincides with the heavy holes' effective potential barrier, whereas for devices with higher Voc, PL activation energies are consistent the electron confinement potential.

CBE sample No	V_E (meV)	Ea (meV)	V_{HH} (meV)	Voc (V)
672	*68*	*72±4*	*55*	*0.71*
711	*102*	*100±5*	*88*	*0.7*
710	109	**66±2**	**62**	0.6
716	96	**63±1**	**64**	0.62
718	130	**95±2**	**95**	0.62

Table 3: Ea is the experimentally determined PL activation energy, V_E is the calculated potential depth of the electron and V_{HH} is the potential depth of the heavy hole for samples displaying a built-in electric field of 32 +/- 1 kV/cm. The open circuit voltage (Voc) is also listed. Note the concordance between Ea and VE for devices with higher Voc (italicized rows). The potential values in bold are those corresponding to the PL activation energy.

DISCUSSION

Results of escape time calculations reported in table 2 suggest that for all samples with built-in electric field in the vicinity of the critical field [6] the light holes escape first. Although under steady illumination and macroscopic timescales identical numbers of electrons and holes are collected in the solar cell, the dynamics of carrier escape rates may lead to charge distributions and band-bendings that are significantly different. The extremely fast escape of light holes (~fsec) creates a dynamic excess of electrons in the well material, which can alter the local electric field, especially if the carrier density is high (i.e. higher light concentration) [12,13]. If the next escaping carrier is the electron (CBE 672 and CBE 711), this would help reduce the charge imbalance. But, if heavy holes are leaving the well potential before electrons (as in devices with lower Voc), one would expect a significant build up of negative charge leading to a pronounced band bending that could affect detrimentally the carrier collection process. The disparity in open circuit voltage identified in these seemingly very similar InP/InAsP MQW solar cells, can be understood as originating from dissimilarities in hole and electrons escape rates [14]. One thus can speculate that the observation originates from an evolution of the band bending associated with the charge imbalance in the well during the sequential escape of carriers, which results in a localized electric field variation and which may significantly alter recombination rates in quantum well devices. By analogy, one would expect that similar mechanisms would be present in other quantum confined systems (e.g. quantum dots). Before concluding, we shall consider two design rules geared towards facilitating the escape and collection of "photogenerated carriers" in quantum confined solar cells. First of all, designs must facilitate the flow of carriers across the i-region trough the presence of a high built-in electric field for an efficient collection and by favoring rapid vertical carrier transport by use of resonant tunneling in closely coupled dots/wells or by favoring very high speed 2D or 1 D electron gas transfer using vertically assembled wells or wires [15]. Second, designs should allow relatively shorter escape times when compared to recombination times in order to limit carrier loss and maximize carrier escape and collection in general.

CONCLUSIONS

Ideally, in quantum-confined solar cells, as in steady state operation, all carriers would escape at the same time. But in reality, due to differences in band offset, effective masses etcetera between carriers, the escape process, as shown here, is sequential. In this study, a strong correlation was found between experimental and theoretical investigations suggesting that the order in which holes and electrons escape from their respective well potentials may have a profound impact on device performance. This is particularly relevant for solar cells where the built in electric field across the quantum confined region is comparable to the critical one [6]. For the InP/InAsP MQW solar cells studied here, stronger Voc and higher performance is warranted by a faster escape of electrons. The origin of this effect is tentatively ascribed to the kinematical time-dependant evolution of the band bending associated with the charge imbalance in the well during the sequential escape of carriers, which results in a localized electric field variation.

REFERENCES

1. W. U. Huynh, J. J. Dittmer, and A. P. Alivisatos, Science, 295, 2002, p 2425.
2. S. A. Mcdonald, G. Zhang, P. W. Cyr, E. J. D. Klem, L. Levina and E. H. Sargent, Nature Materials 4, 2005, p138.
3. K.W.J. Barnham and G. Duggan, J. Appl. Phys. 67, 1990, 7.
4. A. Freundlich, "Multijunction Quantum well solar cell", United States Patents #6,372,980 April 16, 2002, ibid #6,147,296 November 14, 2000.
5. A. Freundlich and A. Alemu, Phys. Statu Sol. c 2005, p 2979
6. I. Serdiukova, C. Monier, M. F. Vilela and A. Freundlich, Appl. Phys. Lett. 74, 1999, 19.
7. A. Freundlich, A. H. Bensaoula, and A. Bensaoula, J. Cryst. Growth 127, 1993, 246.
8. C. Monier, M. F. Vilela, I. Serdiukova, and A. Freundlich, Appl. Phys. Lett. 72, 1998, 1587.
9. A. Alemu, L. Bhusal, A. Freundlich, in PROC. 31ST IEEE Photovoltaic Specialist Conference IEEE Catalog Number: 05CH37608C (2005) p 129
10. A. Alemu, C. Monier, L. Williams and A. Freundlich, Proc. 17th Space Photovoltaic Research and Technology Conference, NASA publications, 2002, p 32.
11. J. Barnes, E. S. M. Tsui, K. W. J. Barnham, C. McFarlane, C. Button and J. S. Roberts, J. Appl. Phys. 81, 1997, p 892.
12. O. Y. Raisky, W. B. Wang, R. R. Alfano and L. Reynolds, Jr., Appl. Phys. Lett. 79, 2001, p 430.
13. S. C. McFarlane, J. Barnes, K. W. J. Barnham, E. S. M. Tsui, C. Button and J. S. Roberts, J. Appl. Phys. 86, 1999, p 5109.
14. A. Alemu, J. A. H. Coaquira and A. Freundlich, J. Appl. Phys. 99, 084506 (2006)
15. A. Freundlich, A. Alemu, S. Bailey, Catalog 31ST IEEE Photovoltaic Specialist Conference IEEE Catalog Number: 05CH37608C – (2005)p137.

Efficient Thin Polymer Solar Cells with Post-Annealing

Shun-Wei Liu[1], Chih-Chien Lee[2], Ping-Tsung Huang[3], Chin-Ti Chen[1], and Juen-Kai Wang[4,5]

[1]Institute of Chemistry, Academia Sinica, Taipei, 11542, Taiwan

[2]Department of Electronic Engineering, National Taiwan University of Science and Technology, Taipei, 106, Taiwan

[3]RiTdisplay Corporation, Hsin-Chu, 303, Taiwan

[4]Center for Condensed Matter Sciences, National Taiwan University, Taipei, 106, Taiwan

[5]Institute of Atomic and Molecular Science, Academia Sinica, Taipei, 106, Taiwan

ABSTRACT

The development of high-performance organic solar cells with low-cost fabrication processes has become one of the most important tasks in the vast endeavors of releasing the world-wide energy demand from fossil fuels. Nowadays, the power-conversion efficiency of polymer solar cells in excess of 5% has been demonstrated, but they involve complicated film formation mechanisms of thick active layers or delicate design of spacer layers. These approaches, therefore, may increase the series resistance of the devices and complicate fabrication procedure. In this report, we present a highly efficient polymer solar cells with a bulk heterojunction layer of poly(3-hexylthiophene):[6,6]-phenyl-C_{61}-butyric acid methylester (P3HT:PCBM) which is annealed at 130°C for 5 min. in a nitrogen environment ($O_2 < 0.1$ ppm and $H_2O < 0.1$ ppm) before cathode deposition. The annealing temperature is much lower and the annealing time is shorter than previous works. The device exhibits conversion efficiency of 4.9%, fill factor of 53 %, and open-circuit voltage of 0.67 V. These values are still comparable with the highest values reported previously. The annealing process is expected to modify the network morphology of the P3HT:PCBM layer. Finally, the thickness of the active layer is reduced to 50 nm which is much thinner than previously reported values, may facilitating the fabrication of tandem photovoltaic structures.

INTRODUCTION

To provide a truly widespread primary energy source, solar cells have been made from many inorganic materials with various device configurations, such as single-crystal, polycrystalline, and amorphous thin-film structures [1-3]. However, these solar energy devices are very expensive for electrical power generation. Organic solar cells have recently attracted much attention due to their advantages such as simple device structure, low-cost materials of tunable band gap, and compatibility with large-area flexible substrates [4-6]. In fact, the energy conversion mechanisms of organic solar cells make use of the transformation of photogenerated excitons into free charge carriers at donor-acceptor heterojunctions. An efficient organic heterojunction device was proposed in 1986 when Tang used two materials with different electron affinities and ionization potentials [4]. At the interface, the resulting potentials are strong and may favor exciton dissociation. Therefore, in this device the excitons should be formed within the diffusion length of the interface, which can contribute to the photocurrent. This approach is mainly limited by the typically small exciton diffusion length in organic materials [4]. To overcome this limitation, bulk heterojunctions (BHJ) cells were developed for various device structures (polymers and small molecules), where the donor and acceptor material are blended together. Light is absorbed in one of the materials and an exciton is created. These excitons dissociated at donor/acceptor interface in any place. Therefore, it is known that the efficiency of organic devices is determined by three processes: light absorption, exciton formation and dissociation, and charge collection [7]. In this letter, we report our studies on the efficient thin BHJ cells with post annealing processes. The device exhibits conversion efficiency of 4.6%, fill factor of 53 %, and open-circuit voltage of 0.67 V.

EXPERIMENT

The bulk-heterojunction solar cells using poly(3-hexylthiophene) (P3HT) as the electron donor and [6,6]-phenyl-C61-butyric acid methylester (PCBM) as the acceptor were fabricated in the device. Notice that P3HT and PCBM were purchased from Rieke Metals and Nano-C. The chemical structure, work-function, and ultraviolet-visible absorption spectrum for P3HT/PCBM are also shown in figure 1. The polymer devices were fabricated by spin-coating (800 rpm) a blend of P3HT:PCBM in 1:0.8 wt/wt ratio (blend solution was used the chlorobenzene), sandwiched a transparent anode (indium-tin oxide; ITO) and low-workfunction metal cathode (Ca/Al). Polymer solar cells were prepared according to the following procedure: The ITO substrate was first cleaned with detergent, then ultrasonicated in acetone and isopropyl alcohol, and subsequently dried in a hot-plate. Poly(3,4-ethylenedioxythiophene):poly(styrenesulfonate) (PEDOT:PSS) (Baytron P standard grade, HC Stark) was spin-cast on top of ITO ($\sim 10\ \Omega/cm^2$) coated glass with a thickness of ~ 75 nm, which the substrate was dried for 5 min at 175 °C in air and then transfer into the a glove box (< 0.1 p.p.m O_2 and H_2O) for fabricated the active layer of P3HT:PCBM. Finally, a 10 nm Ca and a 80 nm Al thick electrodes were deposited on the blend film by thermal evaporation at $\sim 5\times10^{-6}$ torr, defining an active area of 4 mm^2. These devices also were completed with encapsulation in a glove box. For current-voltage (J-V) curves were measured with a keithley 2400 source meter, under illumination from the Xenon lamp (ThermoOriel 150 W solar simulator) with AM1.5G filters. A Newport 818T-10 thermopile detector is used to calibrate the light intensity ~ 100 mW/m^2 for measurement used. All devices were tested in the air environment.

(a)   (b)

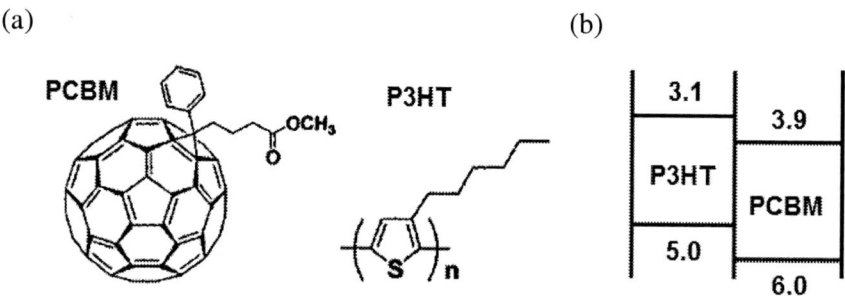

(c)

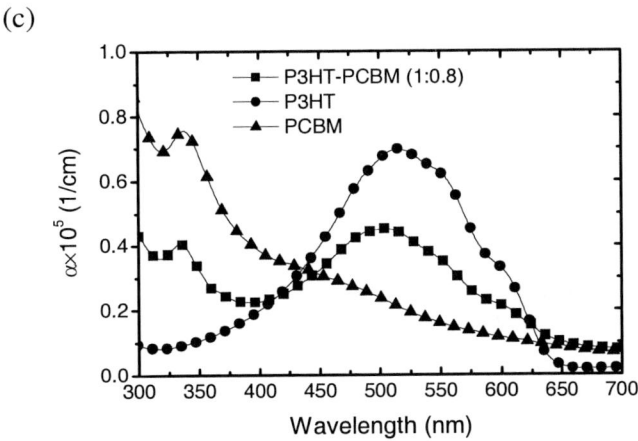

Figure 1. The chemical structure of (a), work-function of (b), and ultraviolet (UV)-visible absorption spectrum of (c) for P3HT/PCBM blend films. The films used in UV measurement were spun-cast on quartz substrate from P3HT/PCBM 1:0.8 wt/wt ratio solution (2.5 mg/ml, chlorobenzene).

DISCUSSION

Figure 2 shows the current-density (J-V) characteristics under AM 1.5 illuminations for different device treatments which included the before-annealing (before cathode deposition) and post-annealing (after cathode deposition). The device performance depends heavily on the annealing conditions as clearly seen from the figure 2. The device with before-annealing shows poor performance with V_{OC} = 0.65 V, J_{SC} = 12.2 mA/cm2, FF = 40 %, and η_{PC} = 3.2 %. After annealing at 150 °C for 15 min or 60 nm, the J-V characteristics have better performance. In addition, after annealing at 150 °C for 60 min, we found that the maximum η_{PC} was obtained to ~ 4.9 %. This value is comparable to the previous report of highest value data for polymer-based solar cells [8].

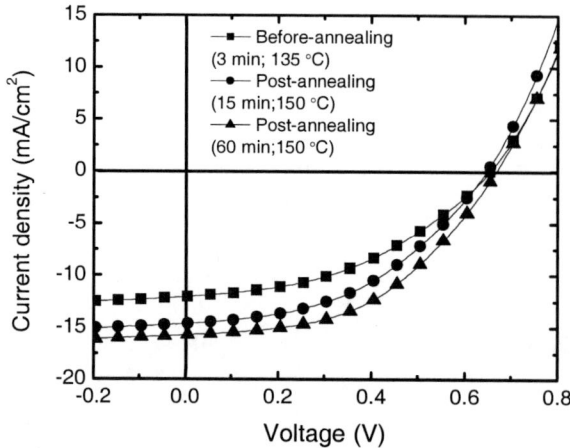

Figure 2. Effect of thermal annealing on the performance of solar cells. The devices were illuminated from a halogen lamp calibrated to an intensity of 100 mW/m².

We attributed that the device with higher efficiency is caused by introducing the post-annealing processes which can modify the surface morphology, improve transport across the interface between the BHJ material and PEDOT:PSS or cathode electrode, and at the same time result in better device performance.

CONCLUSIONS

In this letter, we present the highly efficient polymer solar cells based on bulk heterojunction of poly(3-hexylthiophene): [6,6]-phenyl-C61-butyric acid methylester (P3HT:PCBM) with annealing processes. By applying the post-production annealing at 150 °C (1 hour), polymer solar cells with power-conversion efficiency (η_{PC}) of 4.9 %, fill factor (FF) up to 48 %, and opening circuit voltage (V_{OC}) approaching 0.67 V are demonstrated, which the thickness of active layer was ~ 50 nm. In our experimental results, we found that the improving the net-work morphology of the active layer results in the increased the crystallinity type of the P3HT:PCBM films, indicating in much-improved the device performance.

ACKNOWLEDGMENTS

This work was partially supported by the program promoting University Academic Excellence from Ministry of Education, Taiwan. We are also grateful for support from Academia Sinica and the National Science Council of Taiwan.

REFERENCES

1. M. A. Green, K. Emery, D. L. King, S. Igari, W. Warta, *Prog. Photovoltaics* **13**, 49 (2005).
2. W. Shockley, H. J. Queisser, *J. Appl. Phys.* **32**, 510 (1961).

3. M. A. Green, *"Solar Cells: Operating Principles, Technology and System Applications,"* Prentice-Hall, Englewood Cliffs, NJ (1982).
4. C. W. Tang, *Appl. Phys. Lett.* **48**, 183 (1986).
5. G. Yu, J. Gao, J. C. Hummelen, F. Wudl, A. J. Heeger, *Science* **270**, 1789 (1995).
6. P. Peumans, V. Bulovic, S. R. Forrest, *Appl. Phys. Lett.* **76**, 2650 (2000).
7. P. W. M. Blom, V. D. Mihailetchi, L. J. A. Koster, and D. E. Markov, *Adv. Mater.* **19**, 1551 (2007).
8. G. Li, V. Shrotriya, J. Huang, Y. Yao, T. Moriarty, K. Emery, and Y. Yang, *Nature Materials* **4**, 864 (2005).

Mater. Res. Soc. Symp. Proc. Vol. 1031 © 2008 Materials Research Society   1031-H13-27

Copper Phthalocyanine Nanowire Based Solar Cells

Vijay Singh, Suresh Rajaputra, Sovannary Phok, Goutham Chintakula, and Gayatri Sagi
Department of Electrical & Computer Engineering, University of Kentucky, 453 Anderson Hall, Lexington, KY, 40506-0046

ABSTRACT

Photovoltaic devices based on organic semiconductors are of interest because of their potential as flexible, lightweight and inexpensive devices. One of the promising devices, involves the heterojunction between copper phthalocyanine (CuPc) and 3,4,9,10-perylenetetracarboxylic bis-benzimidazole (PTCBI). Earlier, we reported, the highest V_{oc} (1.125V) in a single organic heterojunction solar cell in an ITO/PEDOT:PSS/CuPc/PTCBI/Al structure. Results were interpreted in terms of a modified CuPc-Al Schottky diode for this thin PTCBI layer case and a CuPc-PTCBI heterojunction for the thick PTCBI case. We also reported the device characteristics of Copper phthalocyanine (CuPc)/Aluminum (Al) Schottky diode solar cells. Here, open circuit voltages (V_{oc}) increased from 220 mV at 15 nm to 907 mV at 140 nm. Analysis of the current-voltage characteristics indicated that tunneling and interface recombination mechanisms are important components of the current transport at the CuPc/Al junction. In this paper, we report the fabrication, materials and electrical characterization of Schottky diode solar cells based on electro-deposited CuPc nanowires. The nanowires were characterized by XRD, UV-Vis absorption spectroscopy, electron microscopy and electrical measurements. Effect of the PEDOT: PSS buffer layer on the nanowire based device characteristics was also investigated.

INTRODUCTION

The development of plastic electronics into a well established technology is a goal currently pursued by many research groups worldwide. The success of plastic electronics depends critically on significant improvements in devices based on organic semiconductors [1,2]. Organic semiconductors like copper phthalocyanine (CuPc) are finding more and more applications in many optoelectronic devices including light emitting diodes [3,4] and solar cells [5-12]. CuPc based solar cells are of interest because of their potential as flexible, lightweight and inexpensive devices. High open circuit voltages have been obtained in Schottky diode solar cells on thermally evaporated films of CuPc [5,6]. However short circuit current densities (J_{sc}) in these cells as in organic semiconductor cells (OSC), in general, are low. The major reason for low J_{sc} in organic semiconductor cells is the small exciton diffusion length of a few nm. Nanowire cell designs [7] offer a way out of this serious limitation and thus a path to high efficiency OSCs. Many promising nanowire fabrication techniques however depend on the technology of electrodepositing organic semiconductors into nanoporous structures like alumina templates. Thus electrodeposition of organic semiconductor films is not only a less expensive technology than deposition by vacuum evaporation, but it is indispensable for fabricating many nanoscale device designs. In this paper, we present the results on the Schottky diode solar cells based on electrodeposited CuPc films with the highest open circuit voltage of 1.190 V reported to date.

EXPERIMENTAL PROCEDURES

Electrodeposition of CuPc on ITO coated glass substrates

ITO coated glass sheets (Delta Technologies, R_s =4-8 Ω/Square) were used as substrates. These sheets were cleaned by sonicating in methanol and acetone and then dried by blowing nitrogen gas. A buffer layer of PEDOT: PSS was spin coated over ITO at 4000 rpm and annealed at 100°C for 90 min in vacuum to improve the adhesion between ITO and buffer layer. CuPc layer was deposited on the samples using two techniques, thermal evaporation and electrodeposition. A 150ml solution of CuPc (0.0040g), 5ml of Trifluoroacetic Acid (TFA) and chloroform is used for the electrodeposition of CuPc. The solution was electrodeposited by supplying a pulse voltage of 120V for 2sec and 0V for 4sec for time duration varying from 2min to 7min. The electrodeposited CuPc samples were annealed in vacuum at 200°C for 6 hours. A 100nm thick Aluminum layer of 0.07cm² was thermally deposited as the electrode. Materials characterization of the films was performed using X-ray diffraction (Bruker-AXS D8 DISCOVER Diffractometer), Optical absorption spectroscopy and Scanning electron microscopy using a Hitachi S-900 FE-SEM. Electrical characterization was performed with an automated I–V tester, solar simulator and an HP 4192A LF impedance analyzer.

CHARACTERISTICS OF ELECTRODEPOSITED CuPc FILMS

Characterization of CuPc with electron microscopy

Electron microscopy images obtained on DC electrodeposited (70V) CuPc films show a needle like morphology with a needle diameter of 90nm (Fig 1). The micrograph shows that these needles are oriented in random directions. When the duration of deposition is increased, there are two layers of CuPc formed. The bottom layer consists of smaller needles and top layer has big needles oriented parallel to the surface. The image also shows that big needles are formed by the accumulation of small needles and have a diameter of about 300nm. The needle like structure of CuPc resembles the β-phase of CuPc. When CuPc was electrodeposited using a pulse voltage, a uniform deposition is observed all over the surface. This also shows a bottom layer of small needles and a top layer of big needles. But the structure, shape and alignment of these big needles is different from that obtained with DC voltage.

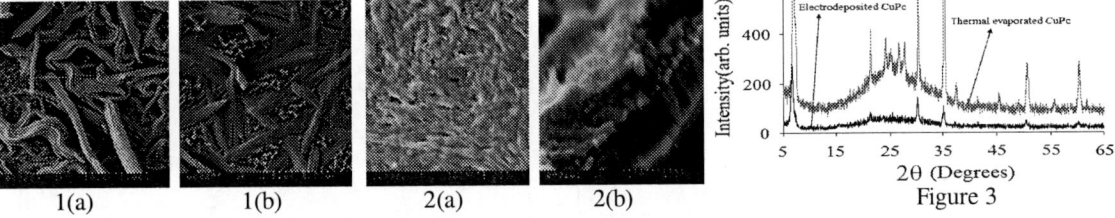

Figure 1. Top view of CuPc layer electrodeposited on ITO through (a) DC voltage (b) pulse voltage; Figure 2: Top and cross sectional view of ITO/PEDOT:PSS/CuPc for the sample electrodeposited with pulse voltage; Figure 3: X-ray patterns showing the comparison of CuPc deposited by two methods.

When PEDOT: PSS is deposited as the buffer layer on ITO, the electrodeposited CuPc shows a completely different surface morphology. As shown in figure 2, the CuPc film does not have any needles on the top layer and shows a uniform surface. This is also different from the particle structure obtained from thermal evaporation. The cross sectional view of the sample shows the PEDOT: PSS layer of 20nm thick and two layers of CuPc. The bottom layer of CuPc has small needles aligned perpendicularly to the ITO surface. These needles are 90nm long and are in close proximity to each other. The top layer is again a uniform layer without any needles. Thus there is a difference in the growth pattern of CuPc films fabricated by different deposition techniques.

Characterization of CuPc using X-Ray Diffraction

The X-Ray diffraction patterns obtained from different methods of deposition of CuPc over ITO/PEDOT:PSS are shown in figure 3. The X-ray diffraction analysis reveals that electrodeposited CuPc films preferably crystallize in β-form; thermal deposited CuPc films crystallize in α-form. Initially CuPc is in β-form (powder precursor for ED), when it is deposited at room temperature. In case of thermal evaporated films, CuPc is in α-form due to the fact that the processing temperature is below 200 ^{0}C. The peaks of ITO substrate are observed at 2θ values of 23^0, 30^0, 35^0, 37^0, 45^0, 51^0 and 60^0 respectively.

Characterization of CuPc using Optical Absorption Spectroscopy

The optical absorption curves of different thicknesses of CuPc are shown in figure 4. Two peaks of absorption obtained both with thermal deposition and electrodeposition occur at same wavelengths i.e. at 620nm and 691nm. The peak at 620nm corresponds to dimer phthalocyanine and the peak at 691nm corresponds to monomer phthalocyanine [10]. As the thickness of CuPc increases the intensity of the peak at 620nm increases stronger than peak at 691nm as shown in figure 4(a). This indicates that percentage of dimer phthalocyanine increases as the thickness of CuPc increases. This shows that the intensity of absorption is less in electrodeposited CuPc films than thermally deposited films. There is a small hump at 350nm for electrodeposited CuPc and a peak at 320nm for a thermally deposited 100 nm thick CuPc as shown in figure 4(b). This shows that CuPc layers deposited with different methods of deposition have different absorption coefficients and wavelengths at which they absorb light.

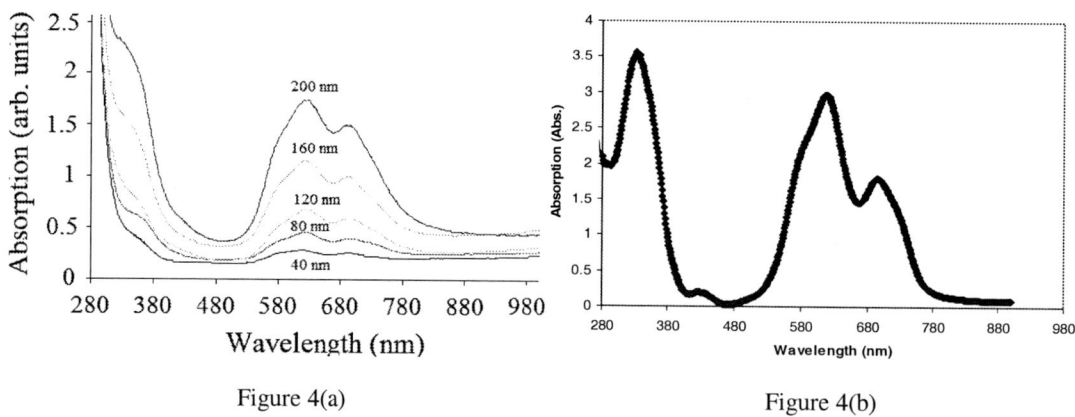

Figure 4(a)

Figure 4(b)

Figure 4: 4(a) Optical absorption spectra for electrodeposited CuPc films of different thicknesses; 4(b) Optical absorption spectra for a 100 nm thick thermal evaporated CuPc film

SCHOTTKY DIODE SOLAR CELLS

In this section, we present the results on Glass/ITO/PEDOT:PSS/CuPc/Al Schottky diode solar cells where the active junction is between Al and CuPc.

Electrodeposited CuPc/ thermally evaporated Al Devices

Figure 5(a) shows the plots of current density versus voltage for an ITO/PEDOT:PSS/CuPc/Al Schottky diode cell in the dark and under illumination; in this cell CuPc film was 200 nm thick and was electrodeposited while aluminum film was 100 nm thick and deposited by vacuum evaporation. A V_{oc} of 1190mV and a J_{sc} of 7µA/cm^2 was observed. An energy band diagram showing a potential barrier of 1.19 eV between aluminum and CuPc is shown in Fig. 5b. High open circuit voltage as well as the low short circuit current density are thought to have their roots in the proposed thin aluminum oxide layer between CuPc and aluminum [3,4].

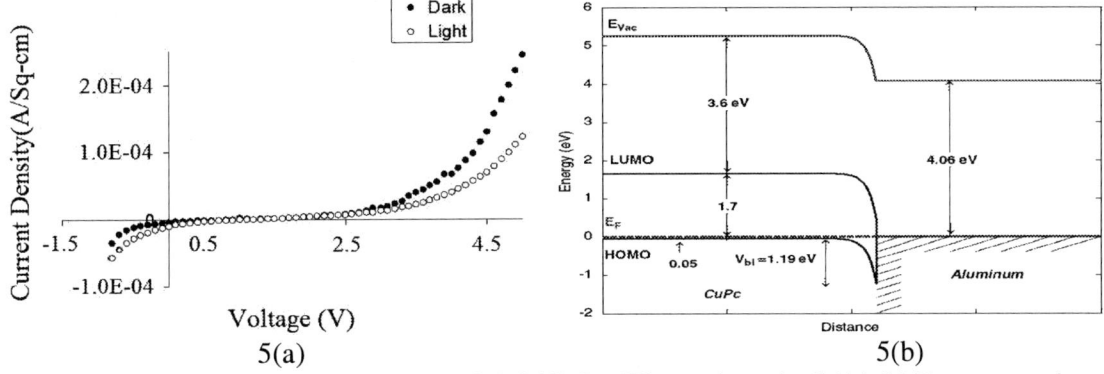

5(a)   5(b)

Figure 5: The dark and light IV curves of ITO/CuPc (Electrodeposited)/Al (b) Representative energy band diagram for the CuPc/Al junction

A set of devices with varying CuPc thicknesses were fabricated and their current voltage characteristics were analyzed to obtain effective diode ideality factors and reverse saturation currents. Results are shown in figures 6,7 and Table 1.

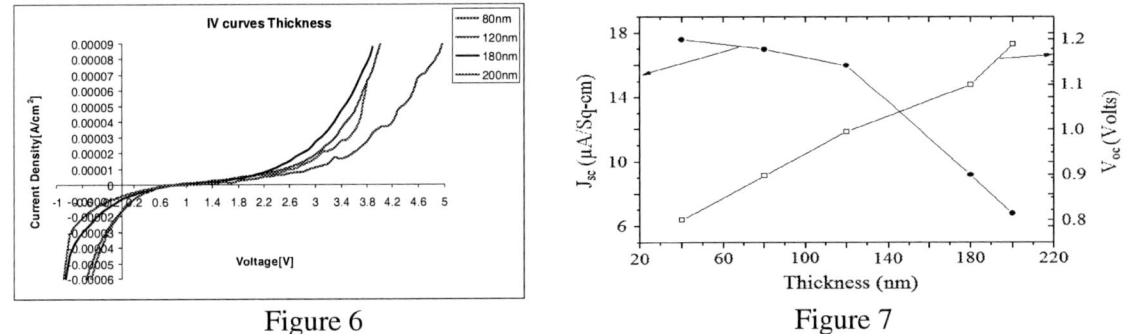

Figure 6   Figure 7

Figure 6: The Light curves for different thicknesses of electrodeposited CuPc for thermally evaporated Al contact; Figure 7: The plot of V_{oc} and J_{sc} versus the varying thickness of CuPc

It was observed that as the thickness of CuPc increases there is an increase in the open circuit voltage V_{oc} and decrease in short circuit current density J_{sc}. This is because the J_{sc} is limited by the thickness and also by other factors like needle like morphology on the top layer and non uniform deposition of CuPc over the ITO.

Thickness of CuPc (nm)	V_{oc}(mV)	J_{sc}(uA/cm^2)	J_o(uA/cm^2)	Series Resistance R_s(KΩ/cm^2)	Ideality Factor(η)
80	898	17	2.18	4.265	24
120	995	16	1.09	1.304	22
180	1100	9.2	48.6	9.564	23
200	1190.7	6.8	35.6	9.011	32

Table 1: Device parameters of the ITO/PEDOT:PSS/CuPc/Al devices under one sun illumination with varying CuPc thickness, when CuPc is electrodeposited and Al is deposited using thermal evaporation.

In our CuPc/Al cells, the measured Ideality factors, η, is larger than 2 and therefore tunneling, recombination-generation currents in the depletion region and recombination through interface states at the CuPc/Al junction are expected to play important roles.

(Electrodeposited CuPc/ electron beam evaporated Al) Devices

Lower open circuit voltages were obtained (as seen in Fig.8 and Table 2) when aluminum electrode was deposited by electron beam evaporation instead of thermal evaporation. We attribute these differences to the penetration of aluminum into the thin CuPc films, when electron beam evaporation is used. This excessive penetration causes shunting paths and reduced voltages.

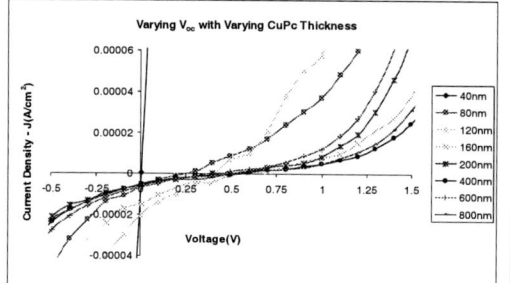

Thickness of CuPc(nm)	V_{oc}(mV)	J_{sc}(A/cm^2)	Series Resistance R_s(Ω/cm^2)	J_{oc}(mA/cm^2)	Ideality Factor
40	98	4.29E-07	1.50	1.0318	18.98
80	397	6.43E-06	5.05	0.0033	13.92
120	493	2.01E-05	5.55	0.0001	12.24
160	599	1.46E-05	2.97	0.0005	13.37
200	599	5.29E-06	7.15	0.0035	17.06
400	599	5.14E-06	17.38	0.0004	12.91
600	494	6.86E-06	9.60	0.0037	16.16
800	494	6.14E-06	5.78	0.1397	24.93

Figure 8   Table 2

Figure 8. The plot of Light curves of different thicknesses of electrodeposited CuPc and e-beam evaporated Al; Table 2: Device parameters of the ITO/PEDOT:PSS/CuPc/Al devices under one sun illumination with varying CuPc thickness, when CuPc is electrodeposited and Al is deposited using e-beam evaporation.

When we compare these results with our results on solar cells with thermally deposited CuPc films published earlier [6], we find that the electrodeposited cells have higher V_{oc} (1.19V). We attribute this to β-phase of CuPc (13, 14), which is dominant in the electrodeposited films as compared to the α-phase in the thermally evaporated CuPc films. The short circuit current density decreases with increasing thickness in the CuPc for electrodeposited cells. In thermally deposited cells the J_{sc} increased slowly with increasing thickness of the CuPc from 15nm to 140nm. In the case of electrodeposited CuPc cells, the J_{sc} decreases as the thickness increases beyond the exciton diffusion length.

CONCLUSIONS

CuPc/Al Schottky diode solar cells made from electrodeposited CuPc films exhibited higher V_{oc} values as the CuPc layer thickness was increased, reaching a value of 1.19 V at a CuPc thickness of 200 nm. This high value of V_{oc} is attributed to the β-phase dominance in electrodeposited CuPc films. J_{sc}, on the other hand decreased with increasing CuPc thickness. The cell performance was found to be critically dependent upon the technique used for Al contact deposition. Much lower open circuit voltages were obtained when the Al was deposited by e-beam evaporation instead of thermal evaporation. This was thought to be caused by excessive penetration of Al into CuPc during e-beam deposition. Diode ideality factor values (η) ranged from 24 to 32. Such high values of η are indicative of the importance of tunneling and interface recombination mechanisms for the current transport at the CuPc/Al junction.

ACKNOWLEDGEMENTS

This work was supported in part by grants from National Science Foundation (NSF-NIRT- ECS-0609064), NSF-EPSCoR (EPS-0447479) and Kentucky Science and Engineering Foundation (KSEF – 148-502-02-27 and KSEF-148-502-03-68).

REFERENCES

1. S. R. Forrest, Nature (London) 428, 911 (2004)
2. F. Yang, M. Shtein, and S. R. Forrest, Nat. Mater. 4, 37 (2005)
3. V.P. Singh, R.S. Singh, B. Parthasarathy, A. Aguilera, J. Anthony and M. Payne *Appl. Phys. Lett.* 86 (2005) 0821061.
4. V.P. Singh, B. Parthasarathy, R.S. Singh, A. Aguilera, J. Anthony and M. Payne, *Sol. Energy. Mater. Sol. Cells* 90 (2006) 798.
5. C.Y. Kwong, A.B. Djurisic, P.C. Chui, L.S.M. Lam, W.K. Chan, *Applied Physics A (Materials Science Processing)*, A77 (2003) 555.
6. Suresh Rajaputra, Subhash Vallurupalli and Vijay P. Singh, *Journal of Materials Science: Materials in Electronics,* 18 (2007) 1147
7. V.P .Singh, R.S. Singh and Karen E Sampson, Chapter VI in "Nanostructured Materials For Solar Energy Conversion" Edited by Tetsuo Soga;   ISBN-13: 978-0-444-52844-5, Elsevier BV (2006)
8. Gayatri Sagi, M.S. Thesis, University of Kentucky, August 2007.
9. Takada, Masaki Yoshioka, Hirokazu; Tada, Hirokazu; Matsushige, Kazumi *Japanese Journal of Applied Physics, Part 2: Letters*, 41, (2002) L73
10. Zhibing He, Gaoling Zhao, and Gaorong Han *phys. Stat. sol. (a)* 203, (2006) 518
11. Peter Peumans, Aharon Yakimov and Stephen R. Forrest, *J. Appl. Phys.*, 93 (2003) 3693.
12. I.G. Hill, J. Schwartz and A. Kahn, *Org. Electronics*, 1 (2000) 5
13. S.I.Shihub, R.D.GouldThis Solid Films 290-291(1996) 390
14. W.Y.Tong, A.B. Djurisic, A.M.C. Ng, W.K.Chan, Thin Solid Films 515 (2007) 5270-5274

Mater. Res. Soc. Symp. Proc. Vol. 1031 © 2008 Materials Research Society 1031-H01-01

Advances in the Research of the Intermediate Band (IB) Solar Cell

Antonio Luque, and Antonio Martí
Instituto de Energia Solar, Universidad Politecnica de Madrid, Ciudad Universitaria, Madrid, 28040, Spain

ABSTRACT

We describe the present state of the intermediate band (IB) solar cell research, a cell concept with very high efficiency potential. A comprehensive presentation of the theory is included followed by a description of its implementation using quantum dots and of the experiments performed to prove the main principles. Presently IB solar cells do not give very high efficiencies; the steps to be taken towards the real achievement of higher efficiencies is described and the use of alloys, instead of nanostructured materials, to fabricate IB cells is also discussed.

INTRODUCTION

The sun is a huge source of energy but it is relatively diluted. Sustainability arguments require that the sun becomes a major source of electricity. For penetrations above 1/3 of the world demand [1] technologies with high efficiency seem necessary. Furthermore such technologies must present a high learning curve [2]. Because of this, the research of solar cells with very high efficiency potential seems important.

Indeed multijunction cells have a very high efficiency potential. Over 40% efficiency has already been achieved [3] under concentrated sunlight with a 3-junction multijunction epitaxial stack of cells. However this stack has about 20 layers and is interconnected by 2 tunnel junctions. Present research in this field is focused in obtaining four junction cells and this is proving to be a hard task. Adding more junctions might be almost impossible and in addition such stacks become highly spectrum sensitive so that it might become unworthy to add many more junctions.

Novel cell concepts, and in particular the Intermediate Band (IB) Solar Cell, may come to the support of the stacks of ordinary solar cells to reduce the number of layers and ease the spectral sensitivity problem and reach an efficiency beyond the 45% that is expected achievable with 4 junctions.

As matter of fact it is interesting to know that for ideal cells, a stack of two IB cells is equivalent to a stack of six ordinary solar cells, and needs a single tunnel junction and not the five that are required for the latter [4].

In this paper we start with a presentation of some selected theoretical aspects of the IB solar cell and then we describe the demonstrators realized using quantum dots (QD). These demonstrators have permitted us to prove experimentally the principles on which the IB solar cell is based, and also identify the problems needed to be solved to render this cell of practical interest.

One of the problems to solve is the weak light absorption by the QD. To overcome this, one of the ways is to develop bulk materials with an IB. This problem is addressed afterwards. Finally conclusions are drawn.

SOME THEORETICAL ASPECTS OF THE IB SOLAR CELL
Increasing the current while preserving the voltage

An IB solar cell [5] is formed of an IB material situated between two ordinary semiconductors —n- and p-type respectively— that play the role of selective contacts to conduction band (CB) and valence band (VB) electrons respectively. As shown in Figure 1, the IB material has a band of states inside the band gap between the CB and the VB. In this way, photons with less energy than the one necessary to pump an electron from the VB to the CB can be absorbed by transitions that pump an electron from the VB to the IB and from the IB to the CB. Thus a full VB→CB electron transition (or electron-hole pair generation) can be completed by means of two photons of energy below the band gap. This mechanism adds to the ordinary one of pumping of electrons from the VB to the CB.

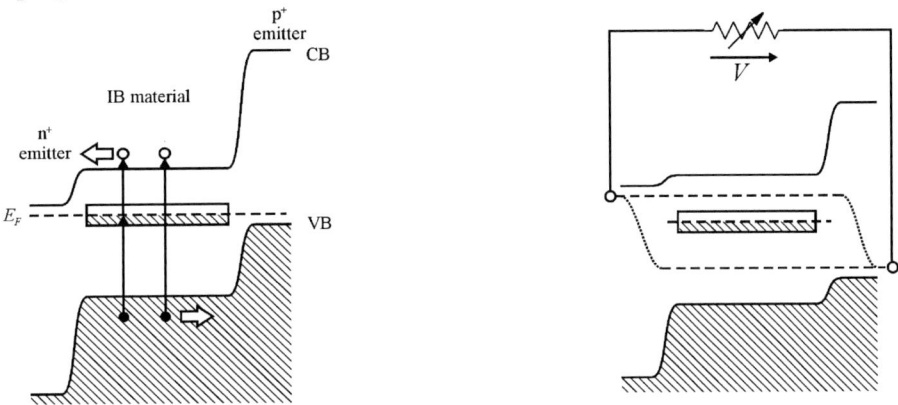

Figure 1. (Left) Photon absorption processes in an intermediate band solar cell. (Right) Out of equilibrium the Fermi level splits into three different quasi-Fermi levels.

Obviously the mechanism described increases the photocurrent because photons of lesser energy are collected. However increasing the current is easy in PV by making the cells in a semiconductor of lower band gap. However this reduces the voltage, that (except when lasing) cannot exceed the bandgap of the semiconductor and the power that can result is reduced (there is an optimum bandgap).

The key point is increasing the current without reducing too much the voltage. For this, three quasi-Fermi levels should appear in the IB material, two of them associated to the VB and to the CB, as in ordinary solar cells, and the third one associated to the IB. The voltage extracted from the cell is precisely the difference of the CB and VB quasi-Fermi levels at the n- and p-contacts respectively (changed of sign and divided by the charge of the electron). Nevertheless, photons of lower energy than this voltage can contribute to the current thanks to the IB.

The limiting efficiency of this concept for maximum concentration (the one providing isotropic illumination on the cell with the radiance of the sun's photosphere) is 63.2% [6] (see Figure 2) to compare with the Shockley-Queisser limit of 40.7% [7, 8] for an ordinary cell in the same conditions.

It has to be understood that in both cases the efficiencies reported refer to the case that any non-radiative recombination is suppressed, the carrier mobilities are infinite, the illumination is an isotropic distributed gas of photons at 6000 K (taken, not accurately, as the photosphere temperature) and the cell lattice is at 300 K To achieve this illumination an ideal concentrator producing a concentration of about 46050 X is necessary[9]. This illumination allows for achieving the ideally highest efficiencies and therefore should be used as the theoretical upper limit of efficiency. However the same efficiency can be achieved by locating the cell in an ideal cavity [10] so that it receives an arbitrarily low irradiance but the radiative recombination is also reduced.

Pure energy transfer between electrons in the IB, by impact ionisation (whose recombination counterpart is an Auger mechanism) or any other mechanism, can also lead to satisfactory IB solar cells [11, 12]. In this mode two sub-bandgap photons pump two electrons from the VB to the IB. Then one of the electrons returns to the valence band and transfers its energy to the other electron that is so pumped to the CB. The limiting efficiency of this cell is somewhat smaller but still very interesting

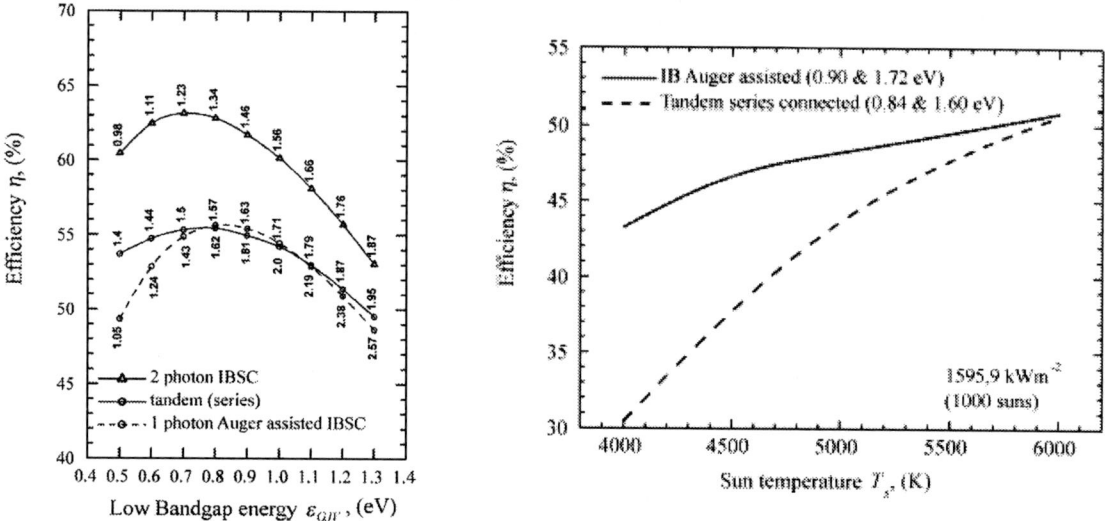

Figure 2. (Left) The limiting efficiency of the IB solar cell vs. the lower bandgap energy (usually IB→CB) in the case of optical VB→IB and IB→CB absorptions (besides the ordinary VB→CB) labelled "2 photon IBSC" and an energy transfer mechanism labelled "1 photon Auger assisted IBSC" and the 2-junction series connected solar cell labelled as "tandem (series)" In all cases the higher band gap has been optimised for maximum overall efficiency. The values in graphic correspond to the value of the larger intermediate bandgap (usually VB→IB). (Right). The limiting efficiency of the IB-Auger-CB solar cell and the limiting efficiency of a tandem of two cells series-connected as a function of the Sun's temperature. In all cases, although the temperature of the sun changes, the input power is kept constant at 1595,9 kWm , equivalent to an irradiance of 1000 suns of a black body sun at 6000 K. The values in brackets correspond to the value of the bandgaps involved.

One very interesting feature of this operation mode is that it is rather insensitive to the spectrum (Figure 2). We have already pointed out the role of the spectral sensitiveness as one of the obstacles to increasing the number of junctions of multijunction cells.

We present in Figure 3 the equivalent circuit of the IB solar cell [13]. This equivalent circuit includes (to the left) three current generators and three diodes representing the recombination associated to radiative recombination between bands (the current represented by this diodes is increased when non-radiative recombination between bands is present). The circuit also shows (square) current generators that represent the transitions associated to the absorption of photons generated inside the cell, through the different radiative recombination processes between bands, and collected by another band. The part at the right consists of the circuital representation of the impact ionization processes.

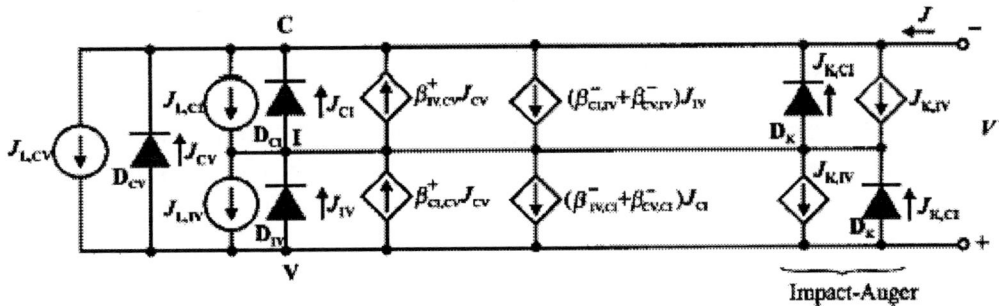

Figure 3. Equivalent circuit of an IB solar cell. In this circuit there are recombination diodes and independent (round) and dependent (square) current generators. The voltage in the C, I, and V points is proportional to the quasi Fermi level positions.

The absorption of the photons emitted in interband transition in a different couple of bands is detrimental for the IB cell performance. This effect is suppressed if each interband (IB→CB, VB→IB, VB→CB) absorption has a spectrum that does not overlap with that of other interband absorptions. The optimum is when they cover the whole spectrum without overlapping [6]. An equivalent condition can be obtained in practice if the three spectrums overlap but there is a large difference in the absorption coefficients in each absorption spectrum in the wavelengths where they overlap [14]. In practice this may require a very effective photon confinement for the weaker absorbed wavelengths.

A very important condition to maintain a high open circuit voltage for a solar cell that uses sub bandgap photons (as in the IB solar cell) is the need to show a voltage above the energy of the lower photon absorbed (of the two involved in the below bandgap absorption). Actually, the absorption of two below bandgap energy photons is a thermodynamic requirement. Several studies have been published illustrating this respect [15-17]. Sometimes thermal escape has been claimed as a mechanism to pump electrons from the intermediate band (or the alternative name given to the level by several authors) to the conduction band. This is indeed possible but if this is the case (see Figure 3), in absence of IB→CB ($J_{L,CI}$) photogeneration and in absence of energy transfer mechanisms (D_K) the thermal escape (the reverse current of the D_{CI} diode) will not happen unless the voltage in point I is above the voltage in point C. This is the same as saying that the IB quasi Fermi level is above the CB quasi Fermi level and implies that the introduction of an IB does not increase but reduces the cell voltage.

Impurity levels vs. bands

A first question arising when IB solar cells are proposed is the following: impurity deep levels are known to introduce a strong Shockley Read Hall (SRH) recombination [18, 19]; why should it be different in an IB cell?

The relatively long lifetimes of electrons and holes in semiconductors are based on the scarcity of particles of energy high enough as to collect the energy delivered in the recombination. Radiative recombination —where these particles are photons— is just a detailed balance counterpart of the light absorption process, which in solar cells is fundamental. In this sense, it is unavoidable, and a solar cell only limited by such recombination, is ideal. On the contrary, non-radiative recombination is in principle avoidable and for ideal performance it must be suppressed

The research about the physics of non-radiative recombination has been a long-standing question in semiconductor physics. Non-radiative recombination can occur by transferring energy to other electrons —Auger recombination— or to phonons. Auger recombination occurs mainly in heavy doped regions such as emitters. For normally doped materials, recombination assisted by phonons is dominant but, still, the low energy of the phonon as compared to the band gap causes that some kind of improbable multiphononic mechanism is necessary for recombination to occur. It is known that impurities leading to deep levels are the cause of the (SRH) recombination.

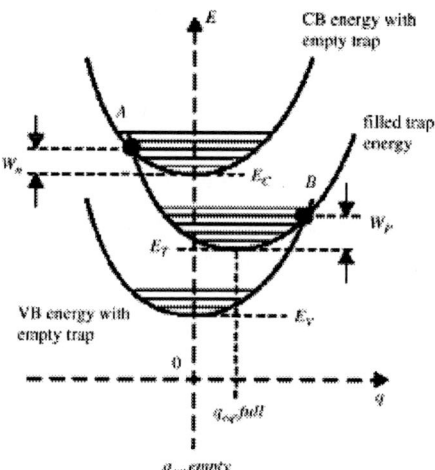

Figure 4. Configuration diagram illustrating the potential energy of the nuclear equation (including electrons and nuclei) per electron along the line of maximum potential slope parameterised by q. The potential energy has the approximate shape $U = \sum_{i,j} b_{i,j} \delta R_i \delta R_j$.

Coordinate q is length along the line of maximum slope in the δR_i space

We believe [20] that the most frequent cause of the bulk non-radiative SRH recombination in the bulk of a semiconductor (at the surface cascaded mechanisms might be dominant) is the so-called Lattice Relaxation Multiple-Phonon Emission mechanism (MPE) [21].

In a semiconductor, the electrons in the bands are characterised by Bloch functions extending across the whole crystal. On the contrary, the electrons in the deep traps are characterised, when the traps are enough diluted, by localized wavefunctions. In this situation, if a transition is produced from an electron in, say, the conduction band to the localized state at the impurity then, there is a large swift charge movement and the charge of an electron formerly distributed across the whole crystal becomes suddenly closely packed around the impurity. This causes the impurity to be very much out of equilibrium with the crystal lattice and consequently it starts vibrating very heavily. A "breathing mode" has been produced.

Precisely, the transition is relatively probable because, in such a non equilibrium position, the full trap level is situated at an energy similar or very close to that of the electrons in the conduction band, as it can be seen in the configuration diagram of Figure 4 (point A), and therefore the transition conserves the energy either strictly speaking, or, at worst, with the emission or absorption of a single phonon.

The violent vibration induced in the impurity by the transition, which can be described by many phonons in the breathing mode, is damped to its thermal value by the successive delivery of these phonons through ordinary phonon-electron interaction. The number of phonons delivered to complete the transition is the so-called Huan Rhys factor that can be in the range of 10 or 12 on many SRH capture processes.

The capture of a hole is similar. The transition (point B in Figure 4) of the localized charge wavefunction in the impurity deep level to the extended state of this electron in the valence band involves a strong charge movement and the impurity becomes again out of equilibrium. A new breathing mode is produced that is damped as before, involving a second Huan Rhys factor.

Increasing the density of traps, so that the Mott transition [22] is produced, will cause the impurity wavefunctions not to be localized anymore. They will therefore become extended. Mott justifies this fact as follows: Removing an electron from the impurity involves leaving the impurity positively charged (at least with respect to the former state) and this implies the appearance of a Coulomb attraction. But this attraction becomes weakened by the presence of other impurity electron wavefunctions that produce screening. This screening leads to a potential energy $-e^2 \exp(-\lambda r)/4\pi\varepsilon_0 r$ (units in SI) where the screening factor λ is given by

$\lambda^2 \cong e^2 N_T / 2\varepsilon\varepsilon_0 kT$. The damping factor increases with the density of impurities N_T.

When the density of impurities is such that $a_H \lambda > 0.888$, $a_H = 4\pi\varepsilon\varepsilon_0\hbar^2 / m_0 e$ being the Bohr radius (no effective mass is used here because the electrons concerned do not move in a band) the Coulomb potential ceases to have a localised state and the electron is free to escape from the impurity. In this case, for all the impurities, an extended state is produced with an eigenfunction that is shared among all the impurities. In the case of a periodic location, this eigenfunction is a Bloch function but this is not necessary in our argumentation.

According with the preceding model, the value for this delocalisation to occur is $N_{Tcrit} = 5.9 \times 10^{19} \text{cm}^{-3}$ at 300 K.

When the delocalisation has occurred, any transition from the conduction band to the impurity band will not involve a large charge movement in the space and therefore, would not imply a strong disequilibrium. In consequence, the impurity electronic energy is not shifted up to the CB energy and the transition between the CB and the deep level cannot occur. In other words, the SRH recombination is prevented.

This actually occurs when the impurities are close enough as to start interacting and forming bands and it ultimately explains why no recombination is to be expected from intermediate bands as those described in the preceding section. It sets a difference between what has been called Impurity Photovoltaics [23] and Intermediate Band Solar Cells.

QUANTUM DOT IB SOLAR CELLS

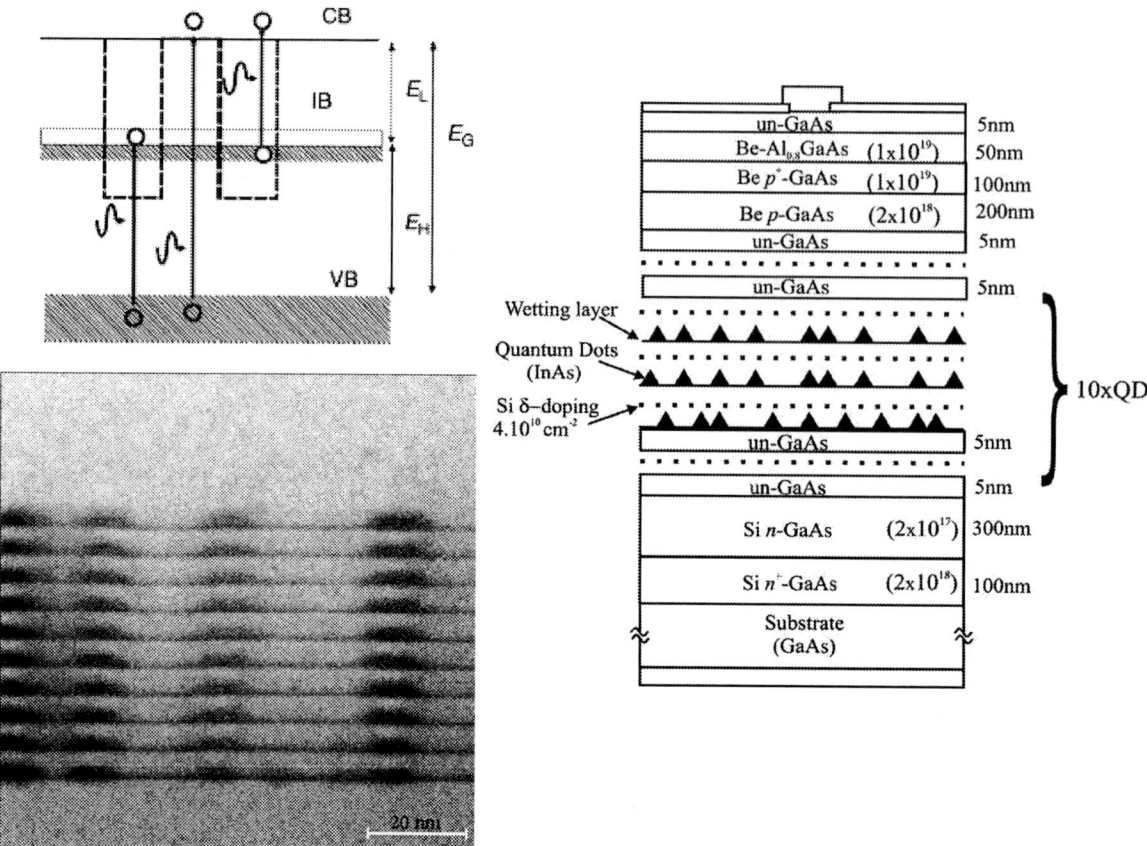

Figure 5. (Upper left) Energy sketch of an intermediate band (IB) material, showing the possible optical transitions. The dashed lines in the bandgap (EG) represent the formation of the IB from an array of quantum dots. CB is conduction band, VB is valence band. (Right) Layer structure of the quantum dot intermediate band solar cells manufactured. (Lower left) TEM photograph of the quantum dot array

Quantum dots have been proposed [24, 25] as one of the means of manufacturing the intermediate band solar cell. In this respect, the intermediate band would arise from the confined electronic states of the electrons in the conduction band's potential wells (see Figure 5).

We promote the use of QDs and not other low dimensional structures like quantum wells [26] or wires because QDs are the structures that have the potential to better isolate the intermediate band from the conduction band by means of a zero density of states.

In order to facilitate the absorption of photons that cause transitions from the intermediate band to the conduction band, the first band has to be half-filled with electrons so that there are enough electrons to be promoted to the conduction band. This can be solved by taking advantage of the modulated doping effect by introducing some n-type doping in the barrier [25] region at an approximate concentration that equals the dot concentration (δ-doping layers in Figure 5).

QD-IBSC prototypes have been manufactured by the University of Glasgow [27] (see Figure 5) within the framework of the collaborative European integrated project FULLSPECTRUM [28]. The basic structure of these cells consisted basically of 10 layers of

InAs/GaAs quantum dots sandwiched by p and n GaAs emitters and is grown by molecular beam epitaxy MBE under the Stranski-Krastanov growth mode. The IB corresponds to the confined level(s) of the quantum dot. This procedure puts the levels rather close in the vertical direction although more spaced and irregularly located in the horizontal one. In the vertical direction they form bands as suggested by the oscillatory shape of the photo-reflectance spectrum due to the QD's that has been observed [29]. The wavefunctions are expected to be localized in filaments that are themselves delocalised. Although some charge localization in filaments occurs, we think that this localization is not strong enough as to be able to lead to a A point of Figure 4 and therefore fundamental non MPE recombination is expected to be prevented, but this conjecture is not properly proven. Localization of wave functions is only detrimental for the MPE recombination. Concerning transport it is not specially detrimental [30], in particular when transport is not blocked in the vertical direction.

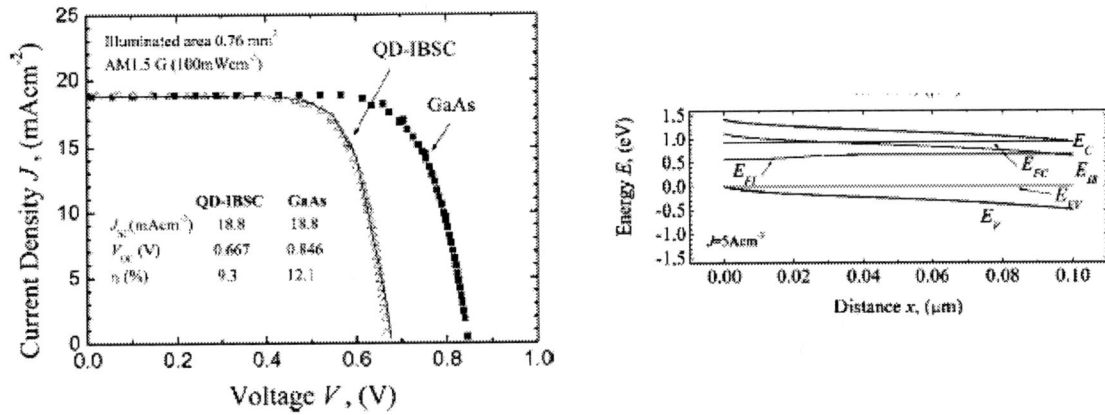

Figure 6. (Left) Current-voltage characteristic of a QD-IBSC and a reference GaAs cell. Solid line corresponds to a fit. The illuminated area consists of the junction area (1.21 mm^2) minus the bus-bar area, but includes the area of the front grid. Current density is referred to the illuminated area. (Right) Modelled bands and VB, IB and CB quasi Fermi levels in this cells at 5 Acm^{-2} direct bias

Figure 6 shows their typical current-voltage characteristic, compared with that of a test GaAs sample possessing the same internal structure as the one shown in Figure 5 but lacking the QDs. As can be seen, the photogenerated current appears to be approximately the same. However, an examination of the spectral photocurrent of the cells (Fig. 7) reveals this is not the case. The QD-IBSC exhibits an extended response for photon energies lower than the bandgap. The contribution to the total current of the cell resulting from below bandgap energy photons is small (1 %) revealing a poor absorption provided by the quantum dots that is unable to overcome the open-circuit voltage loss. This cell has been modelled taking into account that many quantum dots are in a space charge zone and therefore partly filled. In addition the filling of the IB levels have been considered dependent on the illumination and the bias voltage. Fitting has been good for the dark current along five orders of magnitude and also for the IV curve [31]. The degradation of the open-circuit voltage, contrary to what it was expected when formulating the intermediate band concept, can be, at least partially be explained in terms of an effective reduction in the total bandgap of the QD system as a result of the existence of the valence band offset, but it is also likely contributed to by the presence of a higher density of defects in the samples containing the dots.

Experimental proof of the IB solar cell concepts

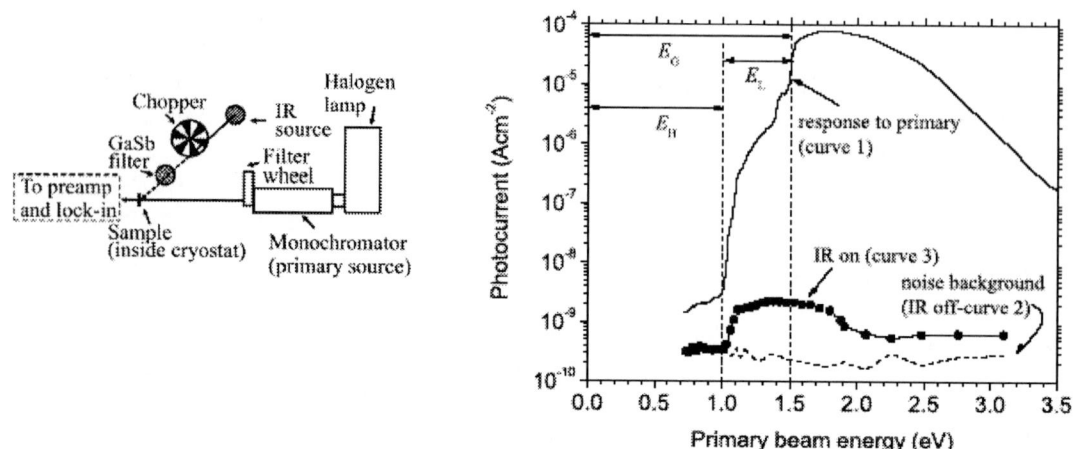

Fig. 7. (Left) Experimental setup used to prove the production of photocurrent as a consequence of the absorption of a photon causing a transition from the IB to the CB. (Right) Photocurrent measurements obtained with the setup at the left [32].

The appearance of a sub-bandgap photocurrent was clearly proven since the fabrication of the first IB solar cells [13] (see curve labeled "response to primary" in Fig. 7. However the doubt existed on whether the IB→CB transition was due to a second photon or on the contrary to a energy transfer of the Impact-Auger type [12] or to a thermal escape that, as it has been discussed is possible but does not allow the voltage preservation, or even to some tunnel assisted punch through of the QDs that by thermodynamic reasons would not have permitted the voltage preservation either. An experiment has been designed and performed [32] that has proven that a two-photon hole-electron pair generation is actually produced. For it, the cell is put at very low temperature (4-30 K) to prevent the thermal escape and the photocurrent produced by a monochromator is recorded. Then an IR broadband light source is filtered so to insure that only IB→CB transitions can be produced by this source and this illumination is chopped (see Fig. 7). A lock-in amplifier collects the photocurrent produced at the chopper frequency only (so the one stimulated by the long wavelength IR source only). A current is produced which is absent if the IR light is switched off. The interpretation is as follows: Above the 1 eV threshold, non-chopped photocurrent always exists because these photons are able to provoke both VB→ IB and IB→CV transitions (although this current could be due as well or solely to the other causes discussed above). However, the chopped current reveals unequivocally that additional current is produced by the low energy photons able only to pump electrons from IB→CB. The second photon is taken from the bias light through the monochromator, and only when these photons have enough energy, above 1 eV, and are not absorbed by the cell emitter, which happens beyond about 1.9 eV, is photocurrent produced at the chopper frequency.

Although the current measured is rather small, the actual photocurrent might be much larger. In reality the chopped current (neglecting second order effects) goes partly by the D_{IV} diode (see Figure 3), lightly forward biased but with small saturation current and by the D_{CI}, reverse-biased but with high saturation current (because the E_{IC} bandgap is small). The result is that most of the chopped generated current is derived by the D_{IV} diode and this current is not measured as an external current.

The electro luminescence of the cells presents a bigger peak that is a signature of the IB→CB transitions and a smaller peak that is the signature of the the CB→VB ones. The ratio of both peaks is associated with the quasi-Fermi levels splitting E_{FC}-E_{FI}. This ratio is variable with the bias proving clearly this statement. This splitting has been measured [33] (in the dark in a direct biased cell) and is of the order of the one shown in Figure 6. This proves the possibility of the separation between quasi Fermi levels that is a condition for preventing a voltage reduction by the increase of the photo-generation spectrum bandwidth.

TOWARDS A PRACTICAL DEVICE

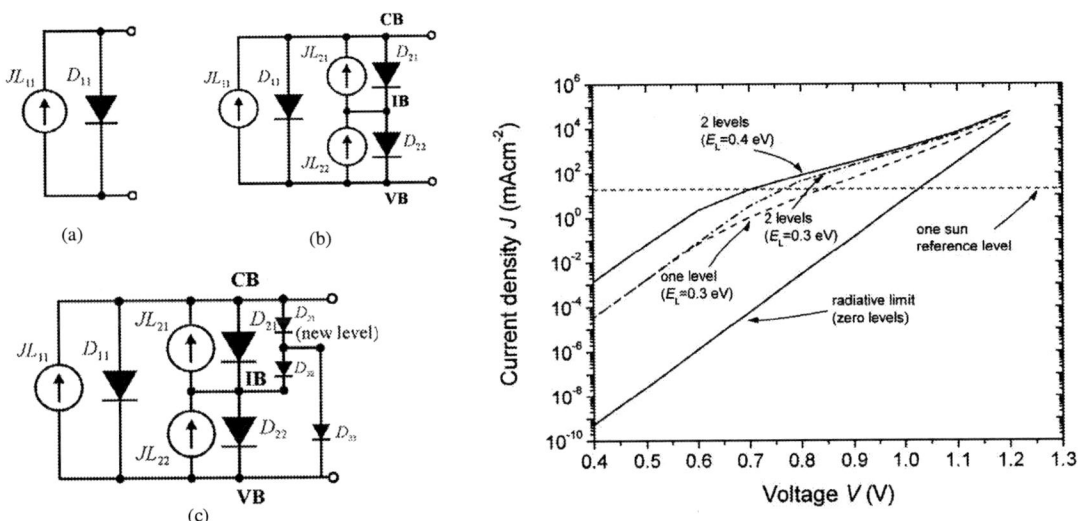

Figure 7. (Left) (a) Simplified equivalent circuit of a single gap solar cell (b) of an intermediate band solar cell (c) of an intermediate band solar cell with additional level appearing in–between the IB and the VB. This new level is assumed not to contribute to the photogeneration of current. (Right) Analysis of the changes produced in the current–voltage characteristic of an IBSC as several levels are introduced in the energy bandgap.

The main reason for reduction of voltage observed in Figure 6 is the reduction of the bandgap associated to the quantum dot fabrication. This reduction is caused by the offset appearing in the valence band (of some 100 meV) and in part by the wetting layer occurring in the QDs grown by the Stransky Kastranov technique that can account for another 60 meV. This almost completes the 200 mV of Voc loss. So it is not adequate to compare a GaAs cell with a GaAs cell with quantum dots. Maybe the comparison might more fair if we make the InAs QD in a matrix of GaAlAs.

It is not sure that this bandgap reduction has to be changed. Our analysis [34] shows that this cell might produce high efficiency if (a) it is operated concentration as are multijunction cells with which these cells should be combined and (b) the absorption of the current is

111

improved. The reason for it appears in Figure 7. In reality all the IB levels (as do many other secondary effects) lose their recombination strength when the cell operates at a high voltage as it is the case in concentrator cells. Note that in this case the dark current of all the cases approaches to the single diode case.

However, the effective absorption of sub-bandgap light is difficult with QD IB cells. The reason is the inherent low density of QDs and the low number of layers so far deposited successfully. In fact our attempts to deposit 50 QD layers instead of 10 have produced poorer cell due to the large density of dislocations induced by the accumulated stresses of the growth and this has increased greatly the recombination at the ordinary semiconductor emitter [35].

The use of strain balanced structures [36] together with advanced light confining structures [37] should lead to significant improvement but much work still remains to be done.

An alternative option is to synthesize an alloy possessing an IB. For some time this was considered impossible. Then ab initio band calculations suggested that certain transition metals, such as Ti, could form suitable IB in III-V [38, 39] and chalcopyrite matrices[40]. In addition the arguments in [20] suggest that many deep level impurities may be precursors of IB materials if the Mott transition is overreached. However the most successful attempt for developing bulk IB materials is based on the band anti-crossing effect [41] according to which the addition of certain isoelectronic impurities with very different ionic radius provokes the detaching of levels that result in an IB situated in the mid of the gap. In this way IB materials have been detected by photo-reflectance spectroscopy in quaternary oxides [42] and nitrides [43]. Unfortunately no solar cells have yet been fabricated, as far as we know, based on this material but the field of bulk IV materials is, in our opinion, very promising

CONCLUSIONS

The IB solar cell promises efficiency limits well above those of any single junction solar cell. They constitute a promising approach to permit the efficiency increase of the present multijunction solar cells when their potential for efficiency increase becomes exhausted. In addition they constitute a research field of high scientific interest as it involves the development of materials with characteristics not existing so far.

IB materials have been developed using the confined states of QDs. This has permitted fabrication of IB solar cells and has permitted us to prove experimentally the basic principles of operation on which this device is based. However, although the use of the cells under light concentration seems to be the way for reducing expected and unexpected reductions of cell voltage, absorbing the light sufficiently with quantum dots is a challenge that so far has not been met.

Promising research has been carried out in the field of alloys exhibiting an IB. Some alloys have been experimentally found based on the concept of band anticrossing and other approaches are in experimentation. However, no solar cell has been realized so far with such materials.

ACKNOWLEDGMENTS

This work has been supported by the European Commission within the project FULLSPECTUM (SES6-CT-2003-502620) and the projects NUMANCIA (S-

0505/ENE/000310) funded by the Comunidad de Madrid and GENESIS-FV (CSD2006-0004) funded by the Spanish National Program

REFERENCES

[1]   T. B. Johansson *et al.*, *Renewable Energy Sources for Fuel and Electricity* (Island Press, Washington DC, 1993).

[2]   A. Luque, Progress in Photovoltaics: Res. Appl. **9**, 303 (2001).

[3]   R. R. King *et al.*, Applied Physics Letters **90** (2007).

[4]   E. Antolín, A. Martí, and A. Luque, in *Proc. of the 21st European Photovoltaic Energy Conference*, edited by J. Poortmans *et al.* (WIP-Renewable Energies, Munich, 2006), pp. 412.

[5]   A. Luque, and A. Martí, Progress in Photovoltaics: Res. Appl. **9**, 73 (2001).

[6]   A. Luque, and A. Martí, Physical Review Letters **78**, 5014 (1997).

[7]   W. Shockley, and H. J. Queisser, Journal of Applied Physics **32**, 510 (1961).

[8]   G. L. Araujo, and A. Marti, Solar Energy Materials and Solar Cells **33**, 213 (1994).

[9]   R. Winston, and W. T. Welford, *Optics of Non Inmaging Concentrators* (Academic, 1979).

[10]   J. C. Miñano, in *Physical Limitations to Photovoltaic Energy Conversion*, edited by A. Luque, and G. L. Araujo (Adam Hilger (IOP), Bristol, 1990).

[11]   S. P. Bremner, C. B. Honsberg, and R. Corkish, in 28th Photovoltaics Specialist Conference (IEEE, 2000), pp. 1200.

[12]   A. Luque, A. Martí, and L. Cuadra, IEEE Transactions on Electron Devices **50**, 447 (2003).

[13]   A. Luque *et al.*, Journal of Applied Physics **96**, 903 (2004).

[14]   L. Cuadra, A. Marti, and A. Luque, IEEE Transactions on Electron Devices **51**, 1002 (2004).

[15]   A. Luque, and A. Martí, Physical Review B **55**, 6994 (1997).

[16]   A. Luque, A. Martí, and L. Cuadra, IEEE Transactions on Electron Devices **48**, 2118 (2001).

[17]   A. Luque, A. Martí, and L. Cuadra, Physica E **14**, 107 (2002).

[18]   R. N. Hall, Physical Review **87**, 387 (1952).

[19]   W. Shockley, and W. T. Read, Physical Review **87**, 835 (1952).

[20]   A. Luque *et al.*, Physica B **382**, 320 (2006).

[21]   D. V. Lang, and C. H. Henry, Physical Review Letters **35**, 1525 (1975).

[22]   N. F. Mott, Rev. Mod. Phys. **40**, 677 (1968).

[23]   M. A. Green, Progress in Photovoltaics: Research and Applications **9**, 137 (2001).

[24]   A. Martí, L. Cuadra, and A. Luque, in *Proc. of the 28th IEEE Photovoltaics Specialists Conference*, edited by IEEENew York, 2000).

[25]   A. Martí, L. Cuadra, and A. Luque, IEEE Transactions on Electron Devices **48**, 2394 (2001).

[26]   K. W. J. Barnham *et al.*, Applied Physics Letters **59**, 135 (1991).

[27]   A. Luque, A. Marti, and A. J. Nozik, MRS Bulletin **32**, 236 (2007).

[28]   A. Luque *et al.*, Solar Energy Materials and Solar Cells **87**, 467 (2005).

[29]   E. Cánovas *et al.*, Thin Solid Films (in the press).

[30]   A. Martí, L. Cuadra, and A. Luque, IEEE Transactions on Electron Devices **49**, 1632 (2002).

[31]   A. Luque *et al.*, Journal of Applied Physics **99**, 094503 (2006).

[32]   A. Marti *et al.*, Physical Review Letters **97**, 247701 (2006).

[33]   A. Luque *et al.*, Applied Physics Letters **87**, 083505 (2005).

[34]   A. Marti *et al.*, Thin Solid Films (in the press).

[35]   A. Marti *et al.*, Applied Physics Letters **90**, 233510 (2007).

[36]   Y. Okada *et al.*, in 20th European Photovoltaics Solar Energy Conference (WIP, Barcelona, 2005), pp. 51.

[37]   A. Luque, Solar Energy Materials **23**, 152 (1991).

[38]   P. Wahnón, and C. Tablero, Physical Review B **65**, 1 (2002).

[39]   P. Palacios *et al.*, Physical Review B (Condensed Matter and Materials Physics) **73**, 085206 (2006).

[40]   P. Palacios *et al.*, Thin Solid Films **515**, 6280 (2007).

[41]   W. Walukiewicz *et al.*, Physical Review Letters **85**, 1552 (2000).

[42]   K. M. Yu *et al.*, Physical Review Letters **91**, 246403 (2003).

[43]   K. M. Yu *et al.*, Applied Physics Letters **88**, 092110 (2006).

AUTHOR INDEX

Alemu, A. 79, 85

Anctil, A. 55, 67

Anthony, R. 25

Bailey, C. 1, 37

Bailey, S. 37

Baumann, A. 49

Buettner, M. 19

Campbell, S. 25

Charlson, J. 31

Chen, C. T. 91

Chintakula, G. 96

Coaquira, J. A. H. 85

Cress, C. 37, 55, 67

Cress, C. D. 1

Deibel, C. 49

DiLeo, R. A. 67

Dyakonov, V. 49

Emmett, K. J. 7

Freundlich, A. 31, 79, 85

Fuhrmann, B. 31

Fujita, Y. 43

Galoppini, E. 61, 73

Hayase, S. 43

Huang, P. T. 91

Hubbard, S. 37

Hubbard, S. M. 1

Inakazu, F. 43

Kashiwa, Y. 43

Kogo, T. 43

Kono, M. 43

Kortshagen, U. 25

Landi, B. 55, 67

Landi, B. J. 1

Lee, C. C. 91

Liu, S. W. 91

Lorrmann, J. 49

Luque, A. 102

Marti, A. 102

Maurer, W. 37

Merrill, A. 55

Noma, Y. 43

O'Donnell, S. 19

Ogomi, Y. 43

Phok, S. 13, 96

Pi, X. 25

Radhakrishnan, G. 31

Raffaelle, R. 37, 55

Raffaelle, R. P. 1, 67

Rajaputra, S. 13, 96

Reinke, P. 19

Robinson, R. 1

Rosenthal, S. J. 7

Sagi, G. 96

Singh, V. 13, 96

Smith, N. J. 7

Taratula, O. 61

Thyagarajan, S. 73

Wang, J. K. 91

Wilt, D. 37

Yamaguchi, Y. 43